W9-AFM-273

Work Horse Handbook

second edition

by Lynn R. Miller

Hartness Library System
Vermont Technical College
One Main Street
Randolph Center, VT 05061

Work Horse Handbook
second edition
Lynn R. Miller
Copyright © 2004 Lynn R. Miller

All rights reserved, including those of translation. This book, or parts thereof, may not be reproduced in any form without the written permission of the author or publisher. Neither the author nor the publisher, by publication of this material, ensure to anyone the use of such material against liability of any kind including infringement of any patent. Inquiries should be addressed to Small Farmer's Journal, PO Box 1627, Sisters, Oregon 97759.

Publisher
Small Farmer's Journal Inc.
PO Box 1627
192 West Barclay Drive
Sisters, Oregon 97759
(541) 549-2064

Authored, illustrated, edited and designed by Lynn R. Miller

Second Edition, First Impression 2004

First printing first edition 1981
Subsequent printings 1983, 1985, 1987, 1988, 1990, 1991, 1992, 1993, 1994,
    1995, 1996, 1997, 1998, 1999, 2000, 2001, 2002, 2003

Library of Congress catalog card number –
ISBN 1-885210-14-0 soft cover
ISBN 1-885210-15-9 hard cover

Also by Lynn R. Miller:

Training Workhorses/Training Teamsters
Horsedrawn Plows & plowing
Haying with Horses
Horsedrawn Tillage Tools
Buying and Setting Up Your Small Farm or Ranch
Why Farm: Selected Essays and Editorials (out of print)
Ten Acres Enough: The Small Farm Dream is Possible
Thought Small: Poems, Prayers, Drawings & Postings
Horses at Work (out of print)
The Complete Barn Book (out of print)

DEDICATION to the first edition

This book is dedicated to my good friends, Queenie, Goldie, Dick, Bud, Flash, Bobbie, Carol, Sarah, Kedde, Major, Bonnie, Betsy, Ted, Tom, Red, Rob, Rowdy, Lil' Joe, Mel, Penny, Roselle, Tatum, Tip, Bob, and Bud. Some of you are still with me, others in new homes and five of you have passed on. All of you have given me purpose. I can never repay you, but I'll keep trying.

DEDICATION to the second edition

In keeping with the first dedication I wish to extend the dedication to additional working equine friends; Cali, Lana, Tuck, Barney, Red, Blue, Rosie, Ben1, Ben 2, Betsy, Teddie, Molly, Polly, Anna, Tommie J, and Abraham the sweetest, most amazing Belgian stallion to have ever sung a snorty note of "come here you babes."

ACKNOWLEDGEMENTS

This book has been a long labor of love. Just as with the first edition, I had to finish it, even though it is not done. Perhaps it never will be. It would not have made it, even this far, without the inspiration, forgiving and help of many people. There have been several dozen truly special teamsters that helped me to "see," they inspired me, most often without knowing it. They should remain anonymous but I thank them here.

And then there are those who have 'given' and 'helped.' Up on top of that list are all the members of my immediate family, with special smiles to my lovely, brilliant and generous wife, Kristi, my sweet young daughter, Scout Gabrielle, and my very special father, Ralph Miller, who has nurtured me so literally without ever pushing. Add to this my grown up children, Justin and Juliet, and my dear departed son, Ian, who all gave so much to allow the first addition to become a reality. And then there are those members of our staff who labored long and hard over this tome: Kathy Blann, Amy Evers and Amy Jo Ferris. All of you, and many more, made this book possible, thank you.

LRM

PREFACE to the first edition

There are pictures around of this writer perched high up on 'Old Jack,' a brown gelding of uncertain ancestry who was dressed for the occasion in his everyday work harness. (Since the photo is a very old one, circa 1918, I was wearing mostly curly hair, short pants and a solemn, if not frightened, expression.) We got rid of our horses a very few years later to my sorrow, and from there on my contacts and use of them were limited to those of my Grandfathers, other relatives and neighbors. I did grow up in an area where the harness horse was in use up until World War II.

A lot of changes occurred as a result of that cataclysm, but aside from the loss of human life, to my mind one of the most devastating was the nearly fatal blow to the use of the harness horse. Right here let me say that I don't really differentiate all that much between the huge draft horses and carriage horses, hackneys or the working ponies. I have a tenderness for them all and I mourned, at least secretly, what I was sure was the certain demise of all those noble animals that had served man so long and well only to be almost literally thrown to the dogs.

When the first stirrings of a revival in such use of horses came to my ears a few years back I hardly dared hope. My delight when it proved true and came close to home was tremendous. I am still moved to be so closely aligned with the revival, even if it is mostly second hand, but then to have even a small part in a book designed to spread the word is a real joy. The sense of awe and bemusement I spoke of is occasioned by my relationship to all this.

In all humility I think I am fitted to introduce the Author; I go back with him to a snowy night in midwinter many years ago, I met him in the arms of the doctor, an almost amorphous bundle a few minutes old and if you think this forward is colored by the pride his father felt at that moment you may be right. As objective as I try to be, his rapid rise to the position he holds in the horse community leaves me with a feeling of wonderment.

I have been right here near him since he first expressed an interest. I helped him find his first team, encouraged him, cautioned him, watched him and still I find it a little incredible. He was listed in an article in the Smithsonian Magazine as one of three Gurus in the work horse revival in America. On the front page of the Wall Street Journal a story on that movement named him as an acknowledged leader and expert in the field. Writings, speeches, lectures, articles in the Small Farmer's Journal and other publications have added to his stature among horsemen everywhere, but, although he seems to thrive on that attention, he is never so happy as when he is working with his horses on the home farm.

In spite of my own lifelong feeling for horses, I can make no claim to have led him into it. When he started working with them, it was on his own and from scratch. As he indicates in the body of this book, good, clear instruction for the beginning or intermediate horseman was not easy to come by. Early in his use of horses he made it a point to seek out older, experienced teamsters. They were of invaluable use, of course, but it was necessary to adapt this advice, varied as it obviously was, to his own practical use. This fact in and of itself gave him insight into the pertinent needs the relative newcomer encounters.

Whatever lack he found in existing writings when he was a novice has been filled in and incorporated along with his own experience and the best of that early advice. In addition to his Editorial hat, he is constantly in contact personally or by letter with people who share his interest or have questions. All this material has been compiled to make it the 'Handbook' that the tyro must have and all others will find helpful.

I don't know how much there can be in predestination or in inherited tendencies for unfulfilled aspirations but along with the harness I started life with a yearning to be involved in the Fine Arts. Although Lynn doesn't make too much of it these days, he did pursue that interest academically and as an avocation up until the time that farming and particularly horses became his dominant pursuit.

Fortunately, publishing and sometimes illustrating a magazine about horses and small farming combines those seemingly varied interests admirably. As the book will make apparent, artistic talent and the eye that accompanies it is a decided advantage in covering a theme of this nature. Not only is he able to diagram and illustrate much of the material presented but the eye trained graphically and to anatomy perceives applications of force, perspectives of design and elements of conformation most conducive to the animal's ability to do the tasks they are bred for. The Farmer and the Artist, far from being antipodes are happily blended to author this work, and from all sides, the practical, the aesthetic, the mundane, the intangible and the interdependence and empathy that has always existed between a man and his equine partners.

Still attempting objectivity, I am struck by the timing of the book. That isn't entirely coincidence, of course; Lynn is totally aware of the demand for such published material because of the swell of interest in the draft horse. Nevertheless, I do see this as a moment that cries out for a definitive work to encourage those who want horses to fill a need in their own lives, to assist those already committed and perhaps to plant a seed in the minds of those only vaguely aware of the possibilities.

The skyrocketing costs of fossil fuel power, the disenchantment with the moil and pollution of mechanically oriented, people-clogged cities, the burgeoning desire to become more self-sufficient, all these things have contributed to the awakening so exciting to those of us who have longed for this day. In addition, there are a growing number of those already farming who are returning to the horse as a practical alternative or a way out of the vicious circle of larger operations with spiraling costs and a constantly dwindling profit margin or outright loss.

None of us expect the horse to replace the combustion powered machines completely but we do see the horse on the way back to a rightful place again. Admittedly the numbers are not great compared to 1914 or in percentage of total population, but put up against 10 or 15 years ago or seen in relation to the wholesale exodus after WW II, the picture is heartening. If this book is instrumental in further turnaround, if it aids those who need help or encouragement, then its primary goal has been achieved. If the present grass roots swell for the return of the harness horse continues, this writer's pleasure will be deep and lasting; if the success and prestige of the book's Author increases by its publication this proud parent's satisfaction will increase proportionally.... And lastly, if you the reader successfully put these precepts into practice, then I'm certain all of us will have occasion to rejoice in the eventual resurgence of man's willing partner, the horse in harness.

Ralph Miller

PREFACE to the second edition

The first has lasted 22 years or more; so possibly he doesn't want to 'change horses.'

I may have used too many superlatives in the previous intros: The Work Horse Handbook on Training Workhorses/ TrainingTeamsters. Time and his extensive study and well founded expertise have proved a fond fathers bragging correct. Nobody is better qualified now to aid and direct the newcomer and those seeking additional facility with the working horse. There are many doing splendid jobs, one-on-one hands on training, but few are ever published.

I hardly qualify as unbiased or any sort of expert on horse handling, but I do believe I know about as well as anyone where the author came from and how he got where he is.

Lynn wasn't born or raised on a farm as his Dad and Grandparents were. He was actually born a half block over the Missouri State Line in the University of Kansas Hospital. That made him a Kansas Jayhawk but so close that he may well have a bit of the Missouri 'show me' attitude in his veins. It's an attitude that has stood him in good stead, especially one starting from scratch.

He was on the thin and disillusioned edge of losing his farm dream when we turned up notice of a selling off of Bob Green's horses at auction, a superlative teamster and owner of championship teams. I believe his attitude on approaching the auction varied between hope and fear; hope for a team and fear that he'd have no success in acquiring what promised to be Green's best show stock. Dare we suggest that the 'Wiz' was still with him for he came home with a team. You know or will realize the feeling especially if you're a complete beginner! And back then, there was no 'Work Horse Handbook.'

When he told me of his success, his grin was as wide as a sunrise over the mountain, but his knees were obviously knocking a tattoo. I've never believed that matched animals necessarily make the best teams, although that may be necessarily in the ring. They <u>were</u> a team, his first one and nearly matched on the eveners, and they were colorful. Queenie a huge red roan and only slightly smaller; Goldie, a palomino hued (blonde) - grade Belgian mare, and like all Bob Green's horses, well broke and experienced; good starting animals for the unschooled, inexperienced new owner.

Again luck was with him and they were patient and willing teachers. For the two or three intervening years until he got his own farm it was obvious that the team was driving Lynn. By trial and error and those patient horses he spent as much time finding his way as he could spare the next three or four years running cattle operations for at least two outfits at three widely scattered locations. Fortunately those absentee owners were willing, perhaps eager to provide room and forage for Queenie and Goldie in exchange for occasional farm chores.

Which came first chicken or egg? Lynn's search for the printed word to augment what he learned by doing as well as watching and questioning experienced teamsters did eventually lead to his decision to write and publish his own. After all, there was comparatively little and much of that very old or suspect if he could find any at all.

Now he sees the need to keep abreast of new and expanded developments and to reflect what he has learned in these last twenty plus years. Understandable surely. My only question is, will he be wanting to do updates in subsequent twenty year periods. If so, someone else will need to be introducing them down the line. At any rate I am grateful to be any small part of it, at least for this far.

Oh! And yes let me paraphrase Dorothy – Lynn, I don't think you are in Kansas anymore!

- Ralph Miller

Introduction to second edition

Today in the year 2003, I work horses.

In harness; for field work on our ranch, for occasional logging, and some highway/parade driving, I work horses. I believe in the practicality of work horses.

For over thirty years, work horses have played a large and important part in the unfolding of my life. They were instrumental in a background which resulted in our creation of the *Small Farmer's Journal*, an international quarterly, featuring, for all of its 28 years, a focus on practical horsefarming. They were also instrumental in my authoring several books on the subject, the *Work Horse Handbook* being the first. For twenty-five years I have conducted auctions and market fairs all centered around the work horse in harness. I have also participated in many working horse clinics, demonstrations and workshops across North America. Horses have been good to me and for me. Beyond that, I know first hand that they **may** be good for, and to you. Further yet, I believe that human society, and the planet, benefits from the values and work which this system manifests. Whenever we do good, beneficial, gainful work at human speed, and in cooperative partnership, the world is better for it.

Working horses in harness is a craft, it's not a science. And not long ago this craft came close to disappearing in the headlong rush to 'progress'. To keep a craft alive it must be practiced, it must belong to practitioners, it must be in use. Having broken a long chain of heritage, we are faced with the need for an "operator's manual" for the craft. Perhaps with such a tool those attracted might actually, more easily, put the system to use. Having started from scratch as a young urban escapee, and experiencing some unnecessary horrors along the way, I am particularly sensitive to the challenges facing anyone who wants to learn this wonderful yet illusive craft.

As with most craft, it is difficult, some might say impossible, to fully capture the teamster's art in words and pictures. All we can hope to do is describe and demonstrate the components of the system and the system itself. We might illustrate the parts, and, though it may be a bit of a reach, we might try to bring others close to the mystery of the craft by metaphor and allegory. We bother with this challenge of teaching because we care. We care about people having the best beginnings and even more than that we care to see horses enjoy safe comfortable working lives. So we involve ourselves in the struggle to keep the teamster's craft alive. And we know we have to do the thing to keep it alive.

As I pen these words, in early 2003, the *Work Horse Handbook* is 24 years old and has been a phenomenal success, having gone through 19 printings going out to nearly 100,000 people. That we would still have occasion to speak of the book is, in my opinion, cause enough for surprise and delight. The original book was conceived with no presumption of lasting importance or value for I had no history, experience, nor credentials to suggest it might be so. The original book has done exceedingly well and continues to this day to sell beyond expectation. To this, I believe, we owe the ongoing presence of a real need for the information. However, I have always felt that the original *Work Horse Handbook* fell far short of its intended goal. That goal was for it to be a comprehensive tool to help people with working the horse in harness while making a strong case for the value of the craft. Herein then, lies a second effort to that end. And, true to the unconsolable artist/word butcher in me, I fear this one also, though more comprehensive, is still far short of a great and good book on the magical, possible world of the work horse.

Beside any issue of whether or not the original book needed improvement, it definitely needed expansion. Over this last quarter of a century, the promised growth in the interest and use of work horses has been realized and in many astounding ways. But speaking solely of the mechanics of the craft: we have been witness to numerous horse saving innovations in tack, machinery, procedure and vet care. The healthier, more comfortable work horse IS the more efficient and willing work horse. And to this end we have many new aids with more on the way.

When the first edition of this book was assembled, back in the 1970's, the art of the teamster was a vanishing craft. Today new practioners, new support industries, new ideas, and lots of young people would appear to guarantee a bright and growing future. It is my sincere wish that this revised book will in some small way continue to encourage increased use, just as the first edition has.

Lynn R. Miller

Bobbie and Carol, two Belgian mares owned by the author. Photo by Nancy Roberts

TABLE OF CONTENTS

The author asking Polly and Anna to back a load of hay. Photo by Kristi Gilman-Miller

CHAPTER ONE

THE WORK HORSE: CONSIDERATIONS FOR THE BEGINNER

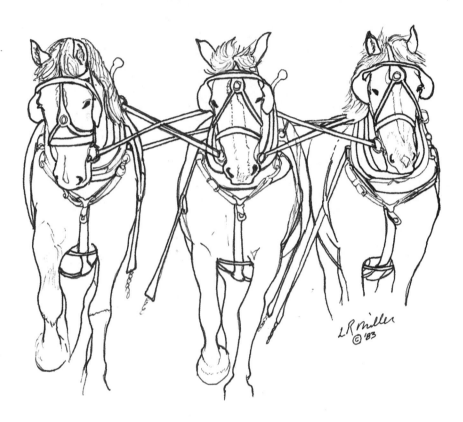

The decision to depend on horses or mules in harness for farm work, logging or highway work is an important one and should not be taken lightly. Aside from romantic notions of involvement in a picturesque scene, most of the considerations are serious. This is not to suggest that the romantic notion is frivolous or in any way detrimental as part of an entire commitment. Quite the contrary; the prevailing motivation behind a majority of good practicing horse farmers today seems to be just that notion. They are attracted to the romance of the system. Whatever the nature of the attraction it must be tempered by consideration of all of the practical questions related to "Why use horses?" Without a careful and clear view of the practical aspects, a person new to the business could very well find himself or herself in either a dangerous, humiliating, confusing or discouraging situation. (It's likely to be a combination of all four.) Make the choice carefully.

learning

The first thing to do is to seek out a good measure of practical information on horses, materials and procedures, as well as advantages and limitations. This book should be but one ingredient of the many which might make up a full education. (There is a directory included in this text which will help in locating some of the places you might go for information pertaining to work horses.) The single most important source of information will be a knowledgeable and experienced human being who you are comfortable with; a person who is doing it or has done it and is willing and capable to communicate his or her knowledge. For most of us, finding that person will mean first finding several people to choose from.

Seek out people who still work horses or who did

it in the recent past and talk to them. As you find people, the "dos and don'ts" of the craft will start to take shape. In the beginning, either you will begin to think that this teamster business is not so complicated or that it is very confusing. It IS complicated and should be confusing at first. This is part of the reason I suggest you find several sources of information in the beginning. Don't settle for "one" source. Stand prepared to hear as many different ways and reasons for working horses as people you talk to. And allow the strong possibility that each and every one of those ideas is correct either in its own way or for the region and job concerned.

From a distance, the accomplished teamster often makes the system look deceptively easier.
photo by Kristi Miller

Allow yourself the benefit of the doubt by talking to several people and asking them, and yourself, questions about the differences in methods and approaches you will hear. In the end you will have to trust your own judgement, so build up a healthy reservoir of information. And then, and only then, select one of your sources to mentor and assist you.

Another little suggestion: Be careful of any information you might gather from a meeting of several practitioners. It might be good, but then again it might not be good. To experience first hand what I suggest; after you think you have a pretty good grasp of a couple of people's horse thoughts, invite them to meet with you and others at some informal gathering to talk horse. You may be surprised to hear the difference: You will witness some amazing about-faces (and maybe even some rather long silences). The best horseman may be ill-equipped to explain, and defend, their approach in a group setting. Whereas pig-headed entrenched pretenders are frequently quick to speak up. And the result may be a slosh of aimless deferrals, or a clash of wills, which will

have little constructive value to your pursuit. So beware or wary of the seminar sources; it seems few want to "look the fool" and the good information often gets set aside.

Hopefully you will discover an intelligent person, with an amazing memory, and a real sensitivity to your need for information, or you will discover a "holdout," someone who is still using horses. Do not, however, disregard someone relatively new at the business. They can make good teachers, because the learning process, with its pitfalls and triumphs, is still fresh to them, and they can communicate this to you.

When I began writing the Work Horse Handbook, back in 1976, there was only one advertised formal workshop where you could go for instruction. If I recall correctly it was at Indian Summer Farm in Cabot, Vermont. Ray Drongesen and I, with occasional help from Doug Hammill, conducted a few workshops in 1978 and 1979. It was in 1980 that I began travelling around North America doing seminars, presentations and workshops, many of which were team or group efforts. Since those early days, dozens of workshops have sprung up all over the continent. Most of them offer exceptional learning experiences. (Many

Horses can be an important ingredient in any deliberate program to improve the self-sufficiency of the farm. Three grey Percherons pull a new I & J cultivator through young Ohio Amish Corn. Photo by Lynn Miller.

The single most important source of information will be a knowledgeable and experienced human being who you are comfortable with; a person who is doing it or has done it and is willing and capable to communicate his or her knowledge. Above, a fitting portrait of Chuck Baley of Pagosa Springs, Colorado and four of his lucky two year old Suffolk students. Chuck also teaches fortunate people. Photo by Lynn Miller.

folks have benefited from the workshop route, while quite a few folks have discovered that the very best teamsters don't always make the best instructors.) There can be considerable expense in this approach towards learning but it is one way, with the right choices, to get the best help while you actually get your hands, safely,

on the lines. I highly recommend this direction. (A list of recommended tutors and workshops appears at the rear of this book.)

Whatever approach you choose, remember to seek out as many sources as possible and accept no "one" article, book, person, workshop, seminar, or

A class picture for a workshop reunion course, 2001, held at Kenny and Renee Russell's Mississippi farm and including Lynn Miller as a co-instructor. The Russell's have conducted many successful workshops and enjoy a well earned reputation as outstanding teachers and human beings.

magazine as representing the "only way." There is no "only way." And when you get confused with the variety I suggest you lean towards "the way" you feel an affinity towards.

In that context, it should be kept in mind that this text is the work of one man and however complete it may seem, it is as limited and biased as the author.

Tom Triplett helps a workshop student understand where and how her hands should work while driving the team. Tom works with his son-in-law Dr. Doug Hammill conducting workshops wherever they are needed. Tom's knowledge of the western teamster traditions is encyclopedic. And his gentle persuasive techniques are part of the reason he is such an effective instructor. Photo by Kristi Gilman-Miller

Bob Olson, of Colorado, helps one of his fortunate students handle the lines for a very tricky unicorn hitch. Photo by Julie Olson.

The author, Lynn Miller, uses his hands to diagram the splice of team lines while Doug Hammill, right, looks on. This photo was taken during the October Work Horse Workshop, 2002, held in Sisters, Oregon. Doug Hammill is a retired equine veterinarian, a western folklorist, associate editor for **Small Farmer's Journal**, *author of the* **Ask a Teamster** *column in the same magazine, and conducts workshops all summer long on his Rocky Mountain Montana Ranch.*
Photo by Kristi Gilman-Miller

Many people, thousands in fact, go each year to Horse Progress Days for its educational aspects. Photo by Lynn Miller

Some workshops specialize in work while others specialize in show driving. Make sure to clarify your needs and wishes and match those to the appropriate teacher and school. In this photo Bob Olson is showing the tricks to driving a six-up in one of his 2002 schools.

difficulties / limitations

If you think you want to work horses, consider these points first:

• It is a complex and subtle business of which the teamster must know everything. If he does not, it may quickly become a dangerous business.

• The all-day working speed of a team of draft horses is between 2 and 4 miles per hour. By how most modern people are accustomed to living, that is slow. It is, for example, slower to work horses than tractors. If speed is important to you or your operation, horses might prove to be a liability. Modern tillage practices and seasonal work schedules on most of today's *factory farms* could not be met satisfactorily by horses. That is not to say that horses cannot get the necessary work done, but it is to say that practices and schedules will have to be modified with the horse in mind. The ideal venue for true horsepower is the small mixed crop and

A photo from the late 1970's of Oregon's Ferd Mantei with his team of Percherons at a plowing competition. Photo by Nancy Roberts. Mr. Mantei was a master horseman who followed the craft out of love.

The older pieces of usable or restorable horsedrawn equipment are getting harder to find. However, over the last twenty years, there has been a dramatic increase in the design and manufacture of outstanding new implements.

livestock general farm which practises the best of rotational cropping and soils management.

- Horses, unlike tractors, must have fuel daily, whether they are working or not. They have to eat. And when working they must be cared for regularly. When the day's field, woods or road work is done, the teamster is not. He must still unharness, feed and curry his horses, looking after anything which needs to be repaired or remedied for the next day's work. A horse farmer must work longer hours, if not as hard, as his tractor counterpart. (That's right, not as hard.)

- In large farming operations (of more than 160 cultivated acres) it may become necessary to bring in outside labor to help drive and care for the horses needed. Finding qualified help can be difficult.

- The supply of well-broke, draft-type horses suitable for work is somewhat limited. The best bring healthy prices now. Good ones are available at fair prices. When supply matches or exceeds demand the price should drop some. Until that time, as demand for horses increases, so also will the price.

- In the first edition of the Work Horse Handbook (1980) I wrote the following paragraph:

Consider the availability of harness and equipment. The recent interest in draft animals has pumped some life into the harness-making trade. Due to some international trade pressures, leather prices are high and so new harness is not cheap. However, with proper care, it remains an excellent investment choice over old harness. Although there are new harness-makers springing into business, demand still appears to exceed supply. Used harness can be found, but many inexperienced people have had unfortu-

nate, discouraging accidents because important parts of used harness broke at a critical moment. The same holds true for old singletrees and doubletrees, as well as neck yokes and poles.

Back then this condition was considered a limiting factor. Today it is an altogether different set of circumstances. How times have changed! As for the availability of harness and equipment, in 1980 we knew of approximately a dozen U.S. harnessmakers. Today, 2003, we lose count at 250. And that does not count Canada, Europe and "Down Under". Several companies are making new harness hardware, collars, sweat pads, singletrees, doubletrees, eveners, neckyokes, poles and sundry hitch gear.

Since 1980 there have been many experiments and innovations in harness construction with synthetic materials making significant inroads. For better or worse some industrial models of supply and manufacture have crept into the harness trade. Today there are small shops which assemble the parts and hardware they receive from larger production facilities. Whereas just 20 years prior, the individual harness maker had to hand cut, sew, rivet, and shape all pieces of each harness.

- Again, under the heading of limitation, in the first edition of the *Work Horse Handbook* I wrote the following paragraph:

Horse-drawn farm machinery is becoming increasingly hard to find in usable condition. Farm machinery companies no longer build this equipment. However, a handful of small shops do still make a few items, such as plows, but on a very limited basis. So limited, in fact, that they are back-ordered and not bothering to advertise.

A Pioneer forecart scaled for draft ponies outfitted with fenders and bench seat. Pioneer Equipment of Dalton, Ohio, is one of the premiere manufacturers of new horsedrawn equipment.

It has all changed for the better. In 1980 there were three to five serious small shops building manure spreaders, plows and forecarts. From that modest beginning we now have more than a dozen serious companies designing and building a whole range of implements including but not limited to; manure spreaders, forecarts (from basic to fully powered), walking plows, riding plows, gang plows, subsoilers, springtooth harrows, spike tooth harrows, a full spectrum of cultivators, planters, sprayers, transplanters, rollers, plastic mulch layers, raised bed formers, round bale handlers, lawn mowers, field mowers, hay rakes, work sleds, and all manner of new vehicles all to be pulled by horses and mules! (See the appendix to this volume for a listing of manufacturers.) Such a vast and wide assortment of excellent equipment has sprung up that an annual field demonstration and trade fair was developed called *Horse Progress Days.* Now ten years and running, these annual doings rotate around to various Amish communities and are held around the Fourth of July weekend. (This author has enjoyed the high honor of assisting for a few years with the announcing chores at this seminal event.)

All indications are that new large and small innovations in equipment and gear are on the horizon. And most of the manufacturers are knowledgeable and committed. In fact it is possible for the inventive and ingenious horsefarmer to augment his or her income substantially through cottage industry aspects of harness and equipment supply. Items as simple as swivel snatch hooks for horse logging or spring loaded line holders have been designed and built by horse farmers for horse farmers. Ripe times, opportunities are definitely there. This would clearly shift harness and equipment supply and support to the advantages column.

Above: A lineup of new I & J Manufacturing horsedrawn row and field cultivators. Implements that did not exist 25 years ago and would not exist today if there weren't an ever growing market for them.

Left: White Horse Machine makes this popular sulky plow. This is only one of the manufacturers of horsedrawn plows enjoying brisk business in the twenty-first century. The best place to view field demonstrations of these new implements and compare one make and model against the others is at the annual Horse Progress Days. See the resource directory in this book for contact and date information.

It is a complex and subtle business of which the safe and successful teamster must know everything.
Bob Nygren plowing with three abreast in 1977.

advantages / opportunities

The major considerations in favor of horse farming include the practical economy, the relative independence, and advantages to soil condition. That might not sound like much after the limitations already discussed but these categories cover a lot of ground.

• If comparing the straight-across cost of setting up a working farm with tractors versus horses, horse farming is less expensive. The difference in cost could be dramatic, depending on the particular combination of variables, i.e., crops grown, technology selected, new tractors or purebred breeding stock versus old tractors or grade horses. All these considerations could affect prices by tens of thousands of dollars. (Please see *Buying and Setting Up Your Small Farm or Ranch* by L.R. Miller.)

• It is also less expensive to operate a farm with true horse-power over tractors. To do an accurate job of accounting actual costs, especially in horse farming,

can be difficult. The aspects are complex. For example, horses produce not only net energy converted to work, but also manure (fertilizer) and offspring. These all need to be calculated into the income production of horses. There are also very subtle, almost intangible, values to be derived from the use of horses as they impose certain limitations, which, if accepted as constructive outlines, will account for the development of individual farm systems which are inherently less costly. For example, pasture, hay and grain are the fuel for horses. Horses can be maintained much of the time on grain stubble and pasture that might be made available on marginal areas of the farm. Areas such as lanes, equipment yard perimeters, wet areas, and fallow fields. And horses work best as a power source if the cropping practices are mixed and divided between spring and fall seasons. Setting up a farm plan to take full advantage of these aspects can reduce out-of-pocket expenses considerably. (A chapter of this book is devoted to the numbers that relate to the cost of using horses.)

If you consider the operations cost for the entire year on the farm, there will be periods when the tractor will not be used and fuel cost will be zero. During those same periods (the horses, because of their tractability, will be able to continue to perform – feeding or spreading manure or yarding logs, etc.) there will be idle days, perhaps even weeks, for the horses and they will have to be fed, but there again with good management, those horses can be on pasture and the cost will be small. In the course of the year, repair work may have to be done to the tractor. Tractor parts and mechanics labor are getting more expensive every day. It is relative to the skills and good fortune of each individual farmer, but often maintenance and repair of the tractor equipment is a major cost item in a farm budget.

Using horses can make a farm less dependent on outside needs such as fuel, oil, grease, mechanical help, tractor parts and such. Horse manure contributes as fertilizer to the farms' soil, reducing outside needs. Horses reproduce, providing their own replacements. Horses, for the most part, are self-repairing. Horses can be an important ingredient in any deliberate program to improve the self-sufficiency of the farm. And horses are good to the soil. They don't compact the soil with a rolling pin effect as tractors do.

All of this seems to suggest that the choice is either tractors or horses. That is most certainly not necessarily so. Many farms enjoy an excellent balanced mixture of tractor and horse-power. These mixed power farms have many advantages to offer. For a variety of reasons many farms may operate most comfortably with a tractor and a team of horses.

There are some hard-to-measure advantages of horses that will be more or less of value to different people. They include the quiet, attractive way of working that horses provide. Anyone who has spent a long, hot, dusty day sitting on a hot, smelly, noisy, vibrating tractor knows how uncomfortable that can be. Horses are quiet and smooth and don't give off 200 degree heat mixed with diesel fumes. In fact, doing the exact same work, horses can be downright comfortable, even relaxing, not to mention rewarding.

No one can guess what will happen in the near future with our energy systems. Today most farms are connected to survival by a gasoline/diesel hose as an umbilical cord. Cut off that hose and the farm grinds to an absolute standstill. It is worth considering as

Gerry Lee and his team of blacks, take a break from the walking plow. Photo by Nancy Roberts.

Horse drawn conveyances are coming back to the cities. Photo by Nancy Roberts.

recent years have proven it is not an altogether impossible situation. In 1980 U.S. fuel prices were little more than half a dollar, today they are at $2 and poised to rocket up. Global uncertainty affects supply and therefore cost.

Horses and horse equipment do not require gasoline or diesel fuels. In light of what has happened in recent years and what could happen again, the horse farmers have good reason to feel healthy in their relative independence and self-sufficiency.

On the tractor farm, the farmer must be concerned about the age and condition of his equipment relative to the work it must do and its resale value. That depreciation scale for farm machinery is a slick, steep, downhill slide. On the other side, horse values, after maturity, hold up well until the horse is too old to work. Add to this the fact that horses are self-renewing. After an initial investment for good work mares, many farms spend no money on replacements. With foals every year to add to the work string, the farmer will find he has horses to sell. Under good management, after the initial investment in horses and equipment is made, the farmer will find himself in an attractive balance of payments situation. With little or no cash outlay for fuel, repairs and maintenance; and no annual debt service on new equipment, and no depreciation in resale values, the farmer has reason to feel healthy. The reduced total overhead makes for a handsome profit margin potential. Horse farming makes money by saving money.

The choice to use horses for log skidding is certainly affected by much of what has been said. Some considerations peculiar to horse logging have to do with the astronomical difference in initial capital investment required between using horses or using machinery. Horses are much cheaper. Also horses do considerably less damage to the forest floor. Horses are a sound choice for ecological reasons. Horse logging is hazardous business with greater inherent risk to horse and driver than most farm work, so great caution is necessary. A good strong horse or team of horses can yard a substantial quantity of logs and should be considered a practical optional power source for logging.

Most people who have considered the practical applications of real horse-power have little difficulty visualizing many of the ways that work horses would fit into the process of farming and logging. That's not always the case when it comes to the use of horses in harness for the transport of goods and people. But in this area of transportation lies perhaps the most dramatic potential for a whole new (and revisited) world of practical applications.

In city traffic (be it large or small), stop lights, stop signs and the rhythm of intersecting streets join with the mass of motor vehicles and pedestrians to set a slow and costly (gasoline-wise) pace. That's not to suggest leisurely – with this "jammed" pace comes tension and anxiety. Within the last thirty plus years, from 1970 on, a few businesses have proven to their own satisfaction that they could haul goods, be it garbage, beer or whatever, and people within cities at a tremendous savings in fuel, with no lost time over customary city travel, with no sanitation or undue public risk hazards, and with tremendous public acceptance. Whether it's self-propelled garbage compacting wagons, beer hauling wagons, omnibuses, furniture vans, parcel delivery service, or carriages – the horse-drawn conveyance within the cities and towns of North America may be making as big a comeback soon.

Those of you under 40 years of age, and some of us over, may not have had any memory, or even concern, about what almost proved to be the extinction of the draft horse in North America and the death of the teamsters' craft.

Following World War II, the decline in draft horse numbers was so rapid that the five major breeds (Belgian, Percheron, Suffolk, Shire and Clydesdale) faced dangerously low breed numbers. Recent years have seen a dramatic popularity turnaround for the big horse. Certainly the much-publicized beer company Clydesdale hitch has done a great deal to reacquaint the general public with the romance of large draft horse hitches as well as to influence some people to put together show and parade hitches of their own. Many draft horse showmen disavow the beer company hitch and take great independent pride in the origin of their own show traditions. But more important have been the rapidly growing number of farmers, ranchers, loggers and homesteaders who have decided for various reasons that they, once again, needed the work horse in harness.

Over these last short ten years the success stories of the new and/or reborn teamsters have fueled more pilgrims. The upshot of it all is a boom in the draft horse market. There simply does not seem to be enough good draft horses and mules to supply the ever-growing market. That, of course, has its good and bad points.

Ten or twelve years ago, early in this draft horse boom, most of the prospective buyers had very little or no practical experience with the craft of working horses in harness. Whether their aim was to skid logs with one horse, feed cattle in the snow with a team, plow fields with six head, cultivate the garden with one mule, pull a manure spreader, haul kids on hay rides, take the buggy to town, pick corn, mow hay or any of the hundreds of things horses can successfully be used for, too many new teamsters had to learn the hard way that this was a craft which looked much easier than it proved to be. Wrecks were so numerous and some so terrible that the only reasons that the interest in work horses didn't become a forgotten fad were; "It was too solid a notion," and "good old human resolve." These pilgrims stuck to their guns and some were successful in finding old-timers who shared some precious, long-ago-taken-for-granted secrets.

Some nice things happened in that process. Some people were rediscovered, respected, needed and made to feel worthwhile. A beautifully subtle craft based on the cooperative communication between man and horse was spared from extinction by being reborn as once again a living legacy. And the fresh youthful outlook of these new teamsters has already accounted for many exciting, and needed, innovations in horse-drawn technology. All of this is to say that the future looks brighter than ever for the marriage of people and work horses.

Yet for the beginner the question remains, "How

Work horses can become part of the family. At Horse Progress Days 2003, an Amish youngster rides his grandfather's Percheron. Photo by Lynn Miller

do I learn what I need to know?" As I said at the beginning of this chapter, hopefully this text will help you learn what you need to know about working horses in harness. But keep in mind that this book is limited because it is 'just a book,' and because the author is 'just one man.' Seek out other sources of information. And be careful to judge the source for what it is.

In this modern, high technology, corporate controlled, "make a fast buck" world there are vultures watching with a keen, well-trained eye for grass roots movements that they can exploit. As soon as the rekindled interest in work horses can be identified by either corporate interests and/or con artists as something marketable, they will do so in droves.

In this digital, out-of-whack world there are so

The author on a feed sled with Cali and Lana. Photo by Kristi Gilman-Miller.

"Working my horses is such a comfort that I feel each experience as if it were a return to home."

few ventures or involvements that allow us to feel really alive and part of the process of energy at work. The great beauty of the notion of working horses is that it gives the individual that special feeling in large measure. It is such an easy process to feel part of and enjoy.

And early in your initiation I sincerely hope that all goes well enough for you to experience a simple question: "Why aren't more people doing this?"

CHAPTER TWO
ATTITUDE AND APPROACH
The Horse as a Thinking, Feeling Mammal

Since the publication of the first edition of Work Horse Handbook, I put together a book entitled *Training Work Horses / Training Teamsters* (see resources at end of this book) which covers in some detail the subjects of horse psychology, man's interface with equine partners, and how one trains a work horse. A core premise, a central idea, is that we may teach our horses and ourselves to be courageous. And on that route we will build a strong smooth reliable working partnership. As horse/human relationships go these days, this is a somewhat unusual approach. However one chooses to approach it, the subject of *'how they are with us and we are with them'* is an important one and cannot be covered in the space of a single chapter of this book. What follows is a severely generalized and simplified introduction to the subject.

I've got more faith in horses than I have in most people. You've probably heard that before but perhaps not in the same context as to which I'm referring.

Horses are honest creatures. If we are allowed a somewhat complete picture of each individual horse's physical world and/or operational structure, (if we may come to know them), it is often easy to measure and understand their motivations, their needs, their perceptions. And, what I mean by honest is that a horse is almost always true to his motivations, his needs, his perceptions: if he wants to eat, if he needs water, if he

Trust and natural finesse make the system work. A young C.J. Shopbell drives two Erskine Shires hitched to a disc harrow. Photo by Heather Erskine.

perceives danger. He is incapable of temporarily setting aside or subverting his motivations to get to some distant goal. This is often mistaken as evidence for a lack of intelligence, a conclusion which says more of human nature than equine smarts. What it means for the horse is that he is almost never lazy, sneaky or deceptive. It is simply not in his nature. People on the other hand are frequently lazy, sneaky and often tripping over themselves in efforts to get an upper hand, or something for free, or a perceived advantage.

Let us back up a few steps and see if these points can be clarified with some observations and a few details:

There are widely held (popular?) beliefs about horses which are unfortunate, unkind, unnecessary, destructive and unintelligent. These sad beliefs include but are not limited to the following:

- *Horses are dumb animals.*
- *Horses are dangerous animals, frequently vicious.*

- *Horses must be forced to cooperate.*
- *Horses must be controlled at all times.*
- *Horses were put here to do our bidding.*
- *Horses have no feelings.*

The horse is **not** "dumb" in any accurate sense of the word. It is **not** true that the horse is by nature vicious and therefore needing to be overwhelmed. It is **not** true that horses are incapable of willing cooperation. It is **not** true that horses must be forced to submit if work is expected.

These myths form the basis for one of the prevalent approaches towards the horse. That approach is grounded in fearful misunderstanding and it's structured around a resulting distrust of the horse. This approach results in an unstable relationship.

Another set of beliefs about horses are unfounded, artificial, sentimental and dangerous. In this vein it is not true that the horse is able to determine good from bad, that the horse is a natural champion and protector of women and children, that the horse is some sort of majestic quivering link with the occult world.

These myths are part of the exploitative imagery of slick magazines, movies and television. An imagery that plays on the hungry vacuum of idle suburban minds. The image of the horse is an easy target for media manipulation. Big, beautiful animal, half wild, pulsing with romance and mystique; doesn't take much to package this image for quick sale to empty heads. And these empty heads too often 'grow up' to join the vast mob of backyard horse owners. These people, when confronted with the reality of the equine, eventually quit with horses or join the ranks of the first group of "initiated" fearful abusers of the horse.

This attitude and approach towards the horse is collectively sustained because most people need their animal(s) to be *wondrous and magical.* Plus they are with their animals part-time and need quick and easy solutions to the problem of actually 'using' their horse(s). With emotion and convenience the primary human concerns, the backyard horse has seldom seen the opportunity to show its true self. It is most telling that when the primary concern with horses is one of *getting the work done*, these magnificent animals almost always show their true selves. Frequently what is identified as a spoiled horse, or a spoiled pet, is actually a wasted horse.

Our history on the continent has been one of greedy exploitation made possible because of abun-

They know you're there, they should be glad you're there. The author, circa mid-seventies, with Belgians Bobbie and Carol getting ready to head out to the field. Photo by Nancy Roberts.

dance. So often you hear, "Why, with so many good horses out there, would you fool with this outlaw animal? Shoot him! Move on. Get a good one."

There is seldom time or concern for what made the animal an *outlaw* (as if that were ever an appropriate term).

what is a horse?

- *A horse is a cognizant, sentient, thinking mammal.*
- *A horses' first line of defense is flight NOT fight.*
- *The horse is a curious, willing, receptive, potential working partner.*
- *Once a deep-seated trust is earned, the horse is reliability incarnate.*
- *We were put here to learn about horses.*
- *The horse is capable of deep emotions.*

confusing respect with fear

The most common attitude, and resulting approach, toward the horse is based on fear (no matter how well-disguised), even though "respect" is the claim. And that is the core of the problem. People are confusing respect with fear. This is a cultural condition. It is dangerous to respect a horse out of fear. Respect is best found to be a result of understanding which includes a measure of trust. And it is difficult to honestly respect a horse without at least a working desire for some understanding of the animal.

It is important to note that, yes, the fear-based controlling approach to horses has resulted, over the centuries, in quite remarkable performance. But it is even more important to point out that, where measurable, the performance demonstrated by partnerships rooted in mutual trust and respect (say even friendship) have won the races, saved the day, plowed the fields, and protected the children. From Bucephalus

All at a human speed and with a human touch. Washington master teamster Clarence Stancil drives
while someone tries out the walking plow handles. This award winning photo by Wendi Ross.

(Alexander the Great's magnificent equine partner) all the way to myriad thousands of champion horses of today, these are the real stuff of legends.

equine senses

In trying to understand the psychology of the horse, we must look to some of the realities of his world. The equine's senses are different. Exactly to what extent this is true is difficult to say, but science, common sense, and observation will tell us that:

hearing

Horses have radar-like acute hearing which serves them in the wild as part of an early warning system. Notice how the horses' ears are directional, aiming at the oncoming sound. They literally turn 180 degrees.

taste

Observing the horse's hesitancy or unwillingness to drink water away from home or eat new feed suggests a strong sense of taste and an ability to differentiate.

touch

The animal's response to gentle handling and petting suggests that there is something communicated through touch and that the animal is sensitive and undeserving, for response, of beating.

smell

The most acute sense of the horse is smell. It, coupled with hearing, works to warn the horse that something's wrong. It is important to understand how sensitive smell is in the horse and to what extent they respond to it.

Here is a quote from an important book entitled *Horse in the Furrow* (by George Ewart Evans. London, Faber & Faber, 1967, 1975) with a story which illustrates the sense of smell.

> 'I heard tell that two carters once called at the Wherry Inn in Halesworth for the usual snack and drink and bait for the horses. They put up the horses in the stable and then went into the pub. After they'd had a couple of drinks one said to the other:
>
> "Shall we have another?"
>
> "No, I reckon we'd better see to the horses."
>
> 'But when they went to the stables they couldn't budge the horses from their stalls. They pulled and they cussed and they swore but the horses wouldn't move an inch. After they tried for a quarter of an hour or so, an old man who happened to be in the yard said to 'em: "What's the matter on 'em? Won't they come out? – I can fix that." He may have been the one who done it –

> mind you, I don't know. But he went inside the pub, and in a minute or two he came out with a jug of milk. He got this jug and put it above the lintel of the stable door and after a minute or two he say: "It's all right: you can take 'em out now." And sure enough, they led the horses out of the door without any trouble at all.' No explanation was given: no explanation was asked for at this stage: for it had become obvious that the horse's acute sense of smell was involved here without question: because milk has the property of absorbing any obnoxious or strong smell arising anywhere near it.'

In this case a smell prevented the horses from moving. The smell was removed and the horses were willing. This story suggests how it is that in some cultures the manipulation of the understood and perceived mind of the horse served to create 'mystiques'. This exists to this day. More often, it is not a manipulation of horses by man so much as it is of people by people; these tricks commonly being used in the sale of horses.

sight

As marvelous and acute as the first four senses are, it is interesting to find how awkward the horses' sight is. The horses' eye has a unique structure and operates differently from most mammals. Rather than a smooth concave retina, the horses' retina is more concave in some places than others and some portions of the lens are nearer the cornea than others. What this means is that the horse can move its head around to focus in on something. They do have muscles, enabling focusing action similar to ours, but they are underdeveloped by comparison. So, if you witness a horse moving its head around in a peculiar fashion it's probably looking for the right combination of lens and light passage to focus on a subject. Because of this strange setup, most horses will shy or bolt away from unexpected and/or unknown sights as there was not time to focus in on the subject and the perception was confused, perhaps even magnified. That is why a small puddle of water or a wind-tossed

Comfort and beauty, power and grace. Charlie Jensen (to the left) and Lorie Jensen (no relation outside of grateful student and deep friendship) check on Lorie's Percherons at an Oregon Draft Horse Association plowing demonstration. Photo by Kristi Gilman-Miller.

The late Marvin Haskell and his horse Rose were both hermits. They lived deep in the Oregon coast woods and at times it was impossible to tell where Marvin ended and Rose began. Photo by Kristi Gilman-Miller.

piece of paper can cause such alarm. Add to this the facts that horses are colorblind and have semi-lateral vision (eyes to the side of the head) and you compound the situation. The semi-lateral vision suggests that often frontal views produce double vision. Imagine if the case were the same for you and you came upon not one, but two enormous black holes, slightly fuzzy ones, that you were asked to walk into!

extrasensory

And then there is the sixth sense. There is all sorts of sophisticated technical evidence from equine research pointing to horses having extrasensory perception. It is not something we can understand yet, but it is something to be mindful of. The author's years of working horses have convinced him that there are many times when horses have read his mind. There is a level of communication possible that is limited only by the human propensity for doubt and suspicion.

Now all this is admittedly a very short oversimpli-

fied introduction into the horses' sensory world, but it's openly meant to suggest that if we could but understand why the horse reacts a certain way perhaps we can earn trust and gain respect.

Maybe a feeling for the limitations of actualities of the horse's sensory world will begin to suggest how it is that the animal looks or thinks about his world and work.

The horse stores experience in the memory bank, probably with complex sensory images rather than visual imagery. In other words, a smell might trigger a memory whereas the sight of something might not. And quite simply, if a horse receives signals, be they smell, sound or whatever, that trigger memories of similar safe and even pleasant experiences, he will actually enjoy falling into the routine asked of him. In reverse, if the experience sends signals that remind them of danger, they will naturally want to avoid or escape from the experience. And when a situation is totally new it will often cause uncertainty and suspicion in the horse because there is no memory to fall back on

John Coffer leans forward as if to help his team up the little incline.

for aid in response. All of this is not too different from how we react to basic physical experiences.

the approach

Your own perception of the horse's condition and capacity should be your guide in how to work with the animal. But if you feel that force is necessary and that you must scare your animal into obeying you – you've got trouble. Or if, on the other side, you are overly protective of your animal, you will not get much done.

For example, if a horse is obviously afraid of something, there are three possible courses of action:

1) Take the horse away from that which frightens him so that he will calm down and behave;

2) Use force or a greater fear to make the horse confront the object, in other words push him past it;

3) Allow the horse time to "figure out" the

object of his fear with you providing security. If the horse can safely view, smell and otherwise experience the object of fear with your reassurances, time will work to either erase or adjust the problem.

If you push, or beat, or frighten, or drag your horse towards that which they originally feared, you've added yourself as a recognizable thing to fear.

Obviously the author favors the third approach. The first approach is, perhaps surprisingly, the worst, as it tends to reinforce in the horse's mind that this object of fear is indeed something to avoid. The second approach will, if repeated, break down any bonds of trust between you and the horse and result in an unstable and unreliable relationship.

Horses enjoy repetition and the security that comes of safe habit. In many farming practices, horses get noticeably frustrated when asked to break from a familiar routine and turn the opposite direction at the end of the field. This is an example of those kinds of moments when the teamster's will is softly tested and the horse tries to discover how much control over the developing relationship he can have. This sort of thing, coupled with apparent judgments made by horses in working processes, indicates that horses are intelligent, thinking mammals.

Horses enjoy company of not only other horses but also people they feel they can trust. Horses demonstrate subtle responses to praise and obvious care and they seem to reward it. There are many cases of horses demonstrating, dramatically, their sorrow at the loss of a companion, be he animal or human. In other words, horses are feeling mammals, capable of the best of what we know to be love.

The best results in working with horses will come from continued efforts to understand the horse's condition and the nature of his communication. Always think about the best ways, the most natural ways, to convince the animal that cooperation is good, even special. Work to find subtle, natural ways to convince the animal that being unwilling to cooperate is bad or generally tedious (not painful or ugly). Find ways to get in the horse's way when it doesn't want to cooperate and find ways to help when he wants to cooperate.

Work to develop the horse as your teammate. Think about the horse as your equivalent and see and feel his limitations as your own. The rewards will be fabulous.

Two year old year old Ashlee Arnold drives Belgian team, Maxine and April.
1998 Photo by Alina Arnold.

CHAPTER THREE
THE DYNAMICS OF DRAFT AND THE MECHANICS OF THE DRAFT HORSE

The title of this chapter might suggest that it is a discussion of harness design. It's not. Outside of passing references to how harness is employed in pulling weight, this is primarily a discussion of the shape and structure of the horse relative to the work he is expected to perform.

The moving parts of a horse are a mass of 'links'; bones, joints, muscles, tendons and ligaments working together to accomplish even the simplest of movements. If we are to depend on a horse in harness to perform certain work on a regular basis, it is essential that his total system be in proper working order. First of all, the animal must be put together properly, with no hereditary or injury-related deficiencies which might prove detrimental to getting the work done. In other words, the horse has got to be put together right or else he is going to have trouble doing the work. Second (and to be discussed in depth later), is the impact of diet and general care on the ability of the horse to work. A fat horse or a starved one will have difficulty performing work.

In the working portion of the horse, potential strength can often be measured in direct relationship to the weakest part of the individual system.

As illustrated on the next page, through the structure of the harness, the horse is able to actually translate a pushing action into a pulling action. Whether it be with a collar or a breast strap harness, the horse's shoulders (or chest) push against an assembly which transfers this into a pulling action with the tugs or traces. Although best communicated through the illustration, perhaps a little additional explanation is in order. Let's limit this to a collar-design harness. The collar, properly fitted around a horse's neck (see HARNESSING chapter), is seated against the shoulder of the horse. Tightly fitted around the collar are steel or wooden 'hames,' which function as a kind of bone structure for the otherwise pliable harness. At a point low on each hame, which corresponds with the wide portion of the collar and the point of the horse's shoulder, a tug or trace is secured. The 'tug' (see *'angle of draft'* below) runs back from the 'hame' alongside the horse and is fastened eventually (directly or indirectly) to the load to be pulled.

As the horse moves forward, he pushes, at the point of his shoulder, against the collar. Through the secured 'hame' assembly and the fastened 'tugs,' that pushing is converted to a pulling action.

If there were no other parts to a harness than the collar, hames, and tugs, and if the load pulled (or the point at which the tugs were fastened) was straight across (or level) this would cause the collar to rock forward (see *'incorrect angle of draft'* on next page) and press, at the bottom, against the horse's windpipe. Even if this were not the case, as the load or 'point of draft' raises higher the pull becomes less natural and more difficult for the horse. The natural slope of the shoulder and the bone structure of the horse's shoulder and front legs (see *'natural shoulder angle'* page 34) lends themselves to a 90 degree (or slightly less) angle of tug to hame (or point of draft – to point of shoulder — to shoulder angle). So there is a natural or optimum angle of draft directly related to the slope of a horse's shoulder, the horse's height and the weight to be pulled.

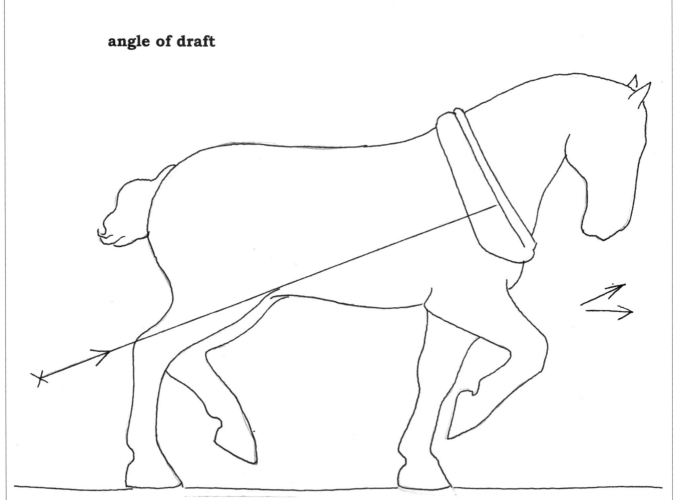

angle of draft

The optimum angle of draft runs from X (which represents the hitching point or where the tug fastens at the evener) to the point of draft at the collar. Through this structure, the horse converts pushing into pulling. Notice that as the horse moves forward there is also a slight up and down movement with each step. The line of draft, in critical applications such as the walking plow, ideally forms a 90° angle with the hames.

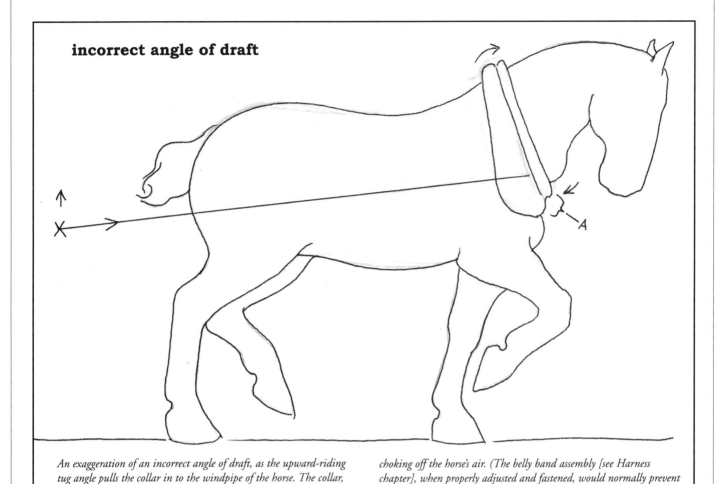

incorrect angle of draft

An exaggeration of an incorrect angle of draft, as the upward-riding tug angle pulls the collar in to the windpipe of the horse. The collar, with this angle of draft, may actually rock forward at the top while choking off the horse's air. (The belly band assembly [see Harness chapter], when properly adjusted and fastened, would normally prevent this action.)

What that suggests is that it is possible for one healthy strong horse to be better shaped to perform work in harness than another equally healthy strong horse. Therefore, it is important to consider the size and shape of any horse expected to work in harness.

All other things being equal, draft or pulling power is in direct proportion to weight, while endurance during work requires the addition of substance (or the size and tone of muscles and bone). Advocates of the old-style Belgian, Percheron, American Cream and Suffolk breeds are more apt to defend bulk and substance in relatively shorter packages as optimum conformation criteria for work horses, whereas Shire and Clydesdale breeders of the British Isles set their standards by another means, using height and bone as their measurement of draft value. As to which is right it is difficult to say, as both arguments have their advantages and disadvantages. I like to think both are justified and certainly the examples bear me out. The short, compact, thickset Suffolk horses have a well-earned reputation as superior farm work animals. But then, the tall, long, heavy-boned Clydesdale has performed his fieldwork admirably for hundreds of years. Certain questions of conformation related to work efficiency are actually best said to be related to personal preference. (See sidebar *"Which Breed?"* page 34)

Two laws of physics suggest that a body in motion tends to remain in motion and that the energy of a moving body is the product of its mass by its velocity. With this the case, the heavy draft horse should have an advantage not only in the ease with which he draws a load when traveling at a rate of speed equal to a lighter rival but also in the ease with which he keeps it going. Gravity overcomes the movement of heavy bodies less rapidly than that of lighter bodies. But don't mistake all of this to suggest that fat is the key to greater efficiency. Fat is soon lost at hard work and is only an incidental assist to pulling. Height and bone as well as muscle tone are the concerns which will

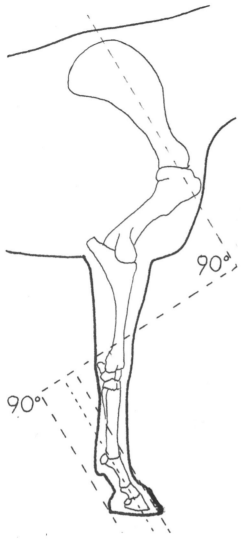

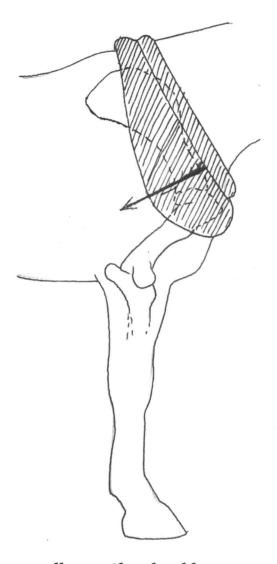

natural shoulder angle

The natural angle of the shoulder and pastern should be the same. When these corresponding angles form ninety degree measures, as illustrated, the 'push-to-pull' action in harness is most efficient and effective. Perhaps even more important is the apparent fact that, all other physiological aspects being equal, these angles translate to greatest comfort for the working horse. The only step we may take to affect these actions, positively or negatively, would be to drastically alter the pastern angle through the angle of the hoof. (see Anatomy chapter).

collar on the shoulder

Illustrating how the padded-collar relates to the skeletal structure of the horse. The point of draft, that being exactly where the tug to hame connection rests on the line of the shoulder, is critical to the working comfort of the horse. If it is too low, a sore shoulder results. If it is too high, it is difficult for the horse to pull. Collars are designed to seat and pad the tug-to-hame connection at their widest part (or the 'draft' of the collar). The 'tug-to-hame' connection is more important than the actual position of the draft of the collar.

Which Breed of Work Horse Do You Recommend?

We are frequently asked to recommend the best breed of work or draft horse. And, for over thirty years we have confidently offered the same answer. Go for the breed that excites you, the one you are drawn towards. You are entering what hopefully will be a lifetime working partnership. If each morning you look out the window and see animals in the pasture and pens which make you smile, perhaps even give you goose bumps of appreciation, you will know you picked the right breed. If you like the look and 'feel' of Norwegian Fjords but go for big black Percherons because you 'think' they are the best breed for the work you wish to do, you may have chosen the wrong breed. If you choose Suffolks or American Creams or Spotted Drafts because they fit the image of the working horse you always dreamed of you've made the right choice. We use Belgians, Percherons, and Belgian/Percheron crosses because we like them all. We believe that a good horse is the best horse.

guarantee a powerful frame. Of course, there must also be concern for the animal's fuel capacity and conversion efficiency. Just as an internal combustion engine needs to have fuel fed regularly and ignited efficiently so it is with a working horse. The animal must have the capacity of converting food efficiently into net energy. The ideal animal combines adequate height, strong ample bone, correct muscle tone, capacity for feed, and the best metabolism possible to aid in feed conversion into net energy.

In an adult horse, little or nothing can be done to affect height and bone. Care and feeding, as well as controlled exercise, can show dramatic changes in muscle, metabolism and energy conversion. In foals, concern about diet and exercise can have effects on the ability of individual animals in reaching their full-inherited potential with regard to bone and height.

The thicker and denser the muscling in a draft

Thicker, shorter, denser muscling is better suited for heavy work in harness.
Photo by Lynn Miller.

horse the better suited that individual will be for heavy work in harness. The longer and thinner the muscling in any individual horse, the greater the potential speed in the animal. A muscle is made up of many cells which join end to end to form a muscle fiber. Muscle

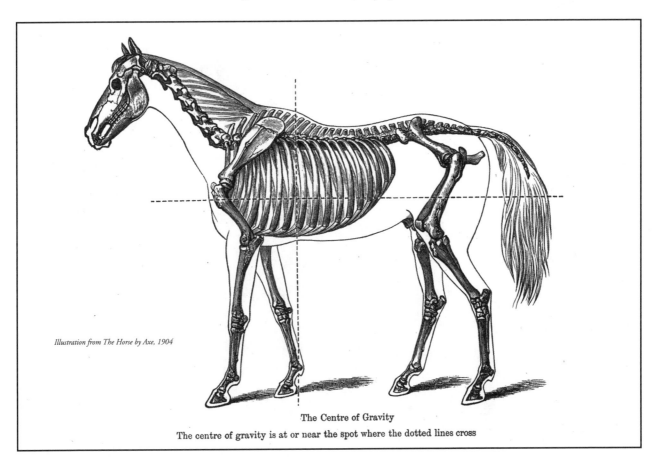

Illustration from The Horse by Axe, 1904

The Centre of Gravity
The centre of gravity is at or near the spot where the dotted lines cross

Above: Jacob Yoder's six black Ohio Percherons demonstrate the muscling and correct hind legs essential to getting the work done. Photo by Lynn Miller.

Left: Beautiful Ardennes and Brabant horses of France retain, under the 'feeding for meat', correctness and power for work. Photo by Jean Christophe Grossettete'

Below: Tom Odegaard and eight North Dakota Belgians walk off comfortably with the big three bottom plow. Photo by Fuller Sheldon.

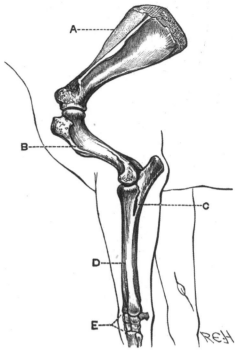

Bones of Left Foreleg

A, Scapula or Shoulder Blade.   B, Humerus or Arm Bone.   c, Ulna.   D, Radius.   E, Carpal Bones, forming the Knee.

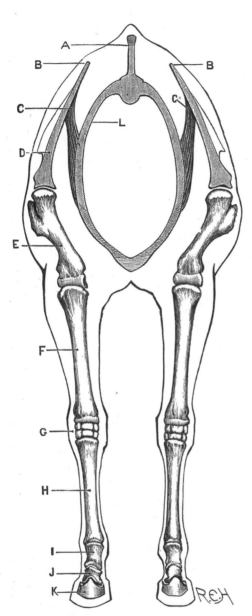

Section through the Chest, showing suspension of the Trunk between the Fore-limbs

A, Vertebral spine. B, B, Scapular cartilages. c, c, Suspending muscles. D, Scapula or blade-bone. E, Humerus or upper arm. F, Radius or lower arm. G, Carpus or knee. H, Large metacarpal bone or canon. I, Os suffraginis or long pastern. J, Os coronæ or coronet bone. K, Os pedis or foot bone. L, Rib.

Illustration from The Horse by Axe, 1904

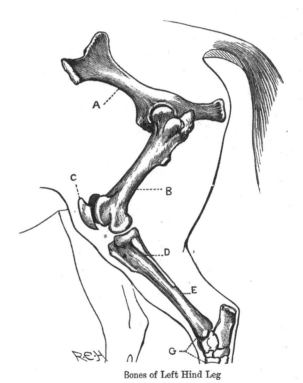

Bones of Left Hind Leg

A, Hip Bone or Pelvis. B, Thigh Bone or Femur. c, Knee-cap or Patella. D, Fibula. E, Tibia. G, Tarsal or Hock Bones.

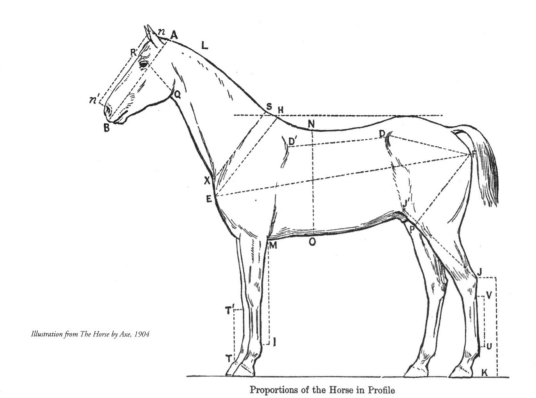

Illustration from The Horse by Axe, 1904

Proportions of the Horse in Profile

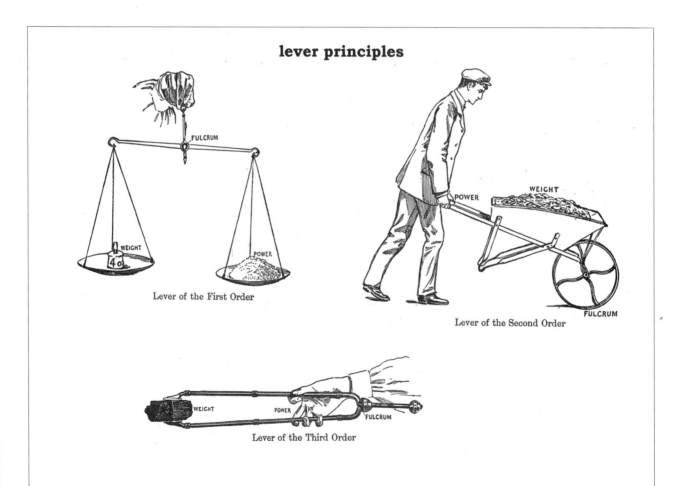

lever principles

Lever of the First Order

Lever of the Second Order

Lever of the Third Order

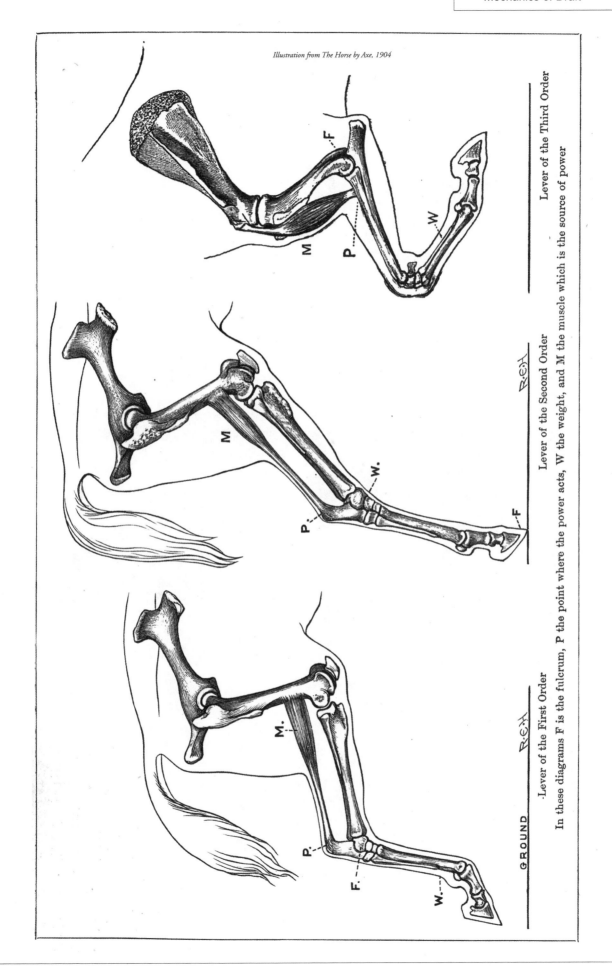

Illustration from *The Horse by Axe, 1904*

Lever of the Third Order

Lever of the Second Order

Lever of the First Order

In these diagrams F is the fulcrum, P the point where the power acts, W the weight, and M the muscle which is the source of power

GROUND

Terry Fitzgerald of Bayfield, Colorado, drives her team of donkeys on a cart. Whatever the breed or type of equine, if you like them, if they light your fire, that's the breed you should work with. Donkeys and small ponies are still being used the world over to get field work done and haul goods. With today's advances we can and should make their working lives as comfortable as possible. The result will be many good years of productivity and shared enjoyment. Photo by Kristi Gilman-Miller.

Raising some of the best Shire horses in the world, John Erskine, of western Washington state, is also a legend having helped many people get started and/or work through problems with their horses. In this lovely photo by Wendi Ross we see a dramatic depiction of angle of draft and point of hitch. In order for that walking plow to work properly, the tugs cannot pull up or push down on the end of the plow beam. Everything must be in a perfect alignment. For in depth information on plowing with horses please refer to the book 'Horsedrawn Plows and plowing', referenced in the resource directory at the rear of this volume.

Larry Livingston plows in the Washington rain with his handsome team of Belgians. Photo by Wendi Ross

Kris Woolhouse of Dexter, Oregon, plows with her two Percherons and a new Pioneer walking plow.

related to contraction rather than size. A muscle cell can contract up to one-fourth of its entire length. After the nerve stimuli have reached the cells (something which occurs almost simultaneously) contraction takes place in about the same length of time, whether the muscle is long or short. A 28-inch muscle will contract its 7 inches in about the same time that a 16-inch muscle will contract its 4 inches. Since the muscular power is applied to the joints at the same place in a draft or large horse as it is in a saddle or light horse, the one with 7-inch muscle contraction cannot help but generate more speed. What we're saying is that

fibers are each independent units of strength yet they run parallel to each other and as a mass form a muscle. A muscle's strength is measured proportionately by the size of its cross section or in direct proportion to the number of fibers that are included. Speed, however, is long muscles generate greater speed than do compact thickset muscles. Length has no relation to strength except in that the greater number of linearly attached cells offers more opportunity for the presence of a weak cell, thus contributing to overall weakness of the

This photo illustrates how the lever principle works: notice that the front legs are used for balance while the back legs are the propulsion. For a short period in the late seventies the author participated in horse pulling competitions. This team, Bob and Bud, did exceedingly well but were retired from the pulling ring in preference of a quiet steady farm field working life. Photo by Matilda Essig.

These six mixed breed horses, all under the tutelage of exceptional Montana horse trainer Bulldog Fraser, are pulling a large spreader across a tilled side hill and demonstrating a more evenly distributed balance as they move ahead.

A young Glenn French and his logging team at a pulling match. As the horses push against their collars, the action is converted to a pull on the traces. Notice the unusual full extension. Photo by Matilda Essig.

Skeeter Thurston of Alberta, Canada mowing with his good team of Belgians. Notice the easy walk and the angle of draft.

Even horses with deformities and infirmities may have the heart and disposition to be excellent work mates.
A team of Belgian mares which belonged to the author. Photo by Nancy Roberts.

muscle fiber. Compact thickset muscles have greater inherent strength than do long muscles. Racehorses will always be long and rangy in type while the best draft horses will exhibit thickness of muscle and stockiness of build.

The skeleton of the horse serves two purposes. One is that it functions as the framework for the body. The other is to provide a mechanical advantage to the horse in motion. This second purpose must be properly understood if you are to be able to identify bone structures which are best suited for work in harness. The bony skeleton of the horse provides various combinations of levers which account for the animal's flexible mobility and strength.

A lever has three working points: the point where power is applied, the point where weight is applied, and the point over which the lever works as a fulcrum or base. There are three classes of levers. The first has power and weight at the ends and the fulcrum in the middle like a seesaw. In the horse this is illustrated when kicking; the foot and lower limb being weight, the power being in the hips and the hock functioning as fulcrum. The second class of lever is when the

fulcrum (or base) is at one end, power at the other end and weight in the middle as with a wheelbarrow. When a horse moves forward you can see this principle demonstrated as the front foot is the fulcrum (or base) the weight comes over and on the joints and the power is in the thighs. This lever is dramatically shown when horses are pulling hard. The strain is in the hocks and the power is in the thighs. The third lever class is where the power is in the middle and the base and weight are at opposite ends, as in a pair of pliers.

The first two lever principles give advantage through increased power. The third lever gives speed at the cost of power. The third class of lever is demonstrated in the horse at the jaw where the weight is at the chewing teeth, the fulcrum at the jaw joint and the power the jaw muscles. Another example would be the limb swinging forward in walking.

The efficiency of the lever depends upon the relative length of the power arm and weight arm. As in a seesaw the longer the power arm the easier it becomes to raise the weight end. So it is necessary to have a longer power arm to gain mechanical advantage. This principle is of great importance to the question of long legged drafters vs. shortset drafters. The weight of the load is a direct force pulling back on, or restraining, the horse. As stated above, the longer the power arm the greater the advantage, hence the theory is that the longer the limb of the horse (which functions as weight arms) the greater the advantage the load will have over the animal. The teamster then requires that his draft horse be lowset in order to gain in pulling efficiency.

Yet there is a second moderating consideration. A horse whose legs are too short may be a disadvantage for in addition to mechanical power the teamster will require a certain speed. A smooth, brisk walk is impossible for a horse after the load drawn has gone beyond a certain point. When that point comes, the pace is a trudging walk at best. This point, at which horses are pulled down to a slow pace, comes at the same time to horses of varying height but equal weight. The horse with the longer swing to his stride will make more rapid progress under identical circumstances than the other fellow. Balancing both principles and affecting the lever by adjusting point of draft (see HITCHING diagrams), each individual conscientious teamster will come to their own happy medium at which the height and substance of the draft horse will be most efficient for both propulsion and progression.

The simple length of the limb is not concern enough if you are after maximum efficiency. The lower portion of the leg functions much like a stilt with muscles being above the hocks or knees. Therefore, the shorter the cannons (see ANATOMY diagram) in relation to the upper leg the greater the mechanical advantage and, as a bonus, the greater the stability and surefootedness.

Another, less understood, aspect of the mechanics of a work horse is what we call "nerve power." It is that ingredient which allows a certain extra staying power or ability to overcome stresses. It is the reserve energy that is somehow called upon to take a heavy load up a steep bank or pull a bogged down wagon out of the mud. It is a quality which some refer to as "heart," others as "honesty," and still others as "strength." No motor has this quality. Whatever it is called it is the quality which has made the horse so easy to love. It is a wide open question whether or not all horses begin life with the same inherent potential for a willingness to give nerve power in harness. The argument narrows some when we discuss the effect man's treatment of the horse can have on the animals "will." I believe that the "nerve" power of a horse is only dependable if the teamster has

"Heart" is that difficult to discern quality which has made the horse so easy to love. Ray Drongesen with Dick, Lucky and Jewell in 1976. Photo by Lynn Miller.

earned the horse's respect and the horse is secure and willing.

The shape, structure and overall condition of the horse has a direct relationship to the ability and willingness with which the animal meets the tasks of pulling a regular load in harness with room for the variations of different human preferences. There is a margin within which the ideal workhorse conformation, constitution and temperament reside. The good teamster will be able to recognize these ideal qualities and select good horses for work. The better teamster will go one better and be able to use harness fit, hitching and even the subtleties of driving to perfect what they easily recognize as an imperfect animal that nevertheless has some qualities so exceptional as warrant saving. How many seemingly flawless horses have lived long enough to prove they were worthless for anything but the show ring? And how many ugly duckling horses, recognized by able teamsters as special, have lived long and fruitful lives as partners in work?

Junior Robbins of North Carolina works his new White Horse 2-way plow with a 3 year old Percheron stallion, a Belgian mare, and a draft mule proving that variety is the spice of life and work! Photo by Paul Breedlove.

CHAPTER FOUR
HORSES, ANATOMY
AND THE FUNDAMENTALS OF CARE

Rutherglen Maestro, 3 year old Clydesdale stallion at an Australian state plow championship at Boyup Brook. Photo by Chas. Holmes.

Climate and diet, varying from country to country, and the uses to which horses were put, have, over the centuries, produced variety in shape, size, color and speed. The earliest civilizations of Europe and Asia began selecting certain superior animals and breeding them with hopes of improving horses for the work in mind. Gradually regions developed families of equine which exhibited consistent characteristics of shape, color, even performance. So in this way began the different breeds of horses.

The two predominant forces in the development of breeds of horses were environment and man's expectations. Environmental effects can be seen when we look at the Arabian horse and think of hot, dry, barren expanses of North Africa where the breed was born. Look, in comparison, to the Fjord horse of Norway or the Clydesdale horse of Scotland. Hot and barren, cold with short growing season, or temperate and lush; all have their marked influence on the indigenous breeds of livestock. Add to this the different expectations of man, the domesticator of the horse. In North Africa, the nomadic Bedouin tribes expected the horse for transportation of rider/soldier with speed, heat tolerance, and easy keeping qualities. So they selected the best animals to suit their needs. In Scotland and all of the British Isles long ago the great horse had been selected, because

GOLDEN KNIGHT 15447 (15581)

of its size and bulk, to carry the knights in armor to battle. After the Crusades of the Middle Ages the farmers of the British Isles continued to breed the great horse for height, bone and substance but for different purposes. These horsemen expected horses to pull plows and haul heavy loads.

Thus there were developed in different regions horses of different breeds. Some desired horses for heavy work, animals of heavy body, stout limbs, and strong muscles. Others desired horse for speed, animals of lighter frame, smaller

Brabant Mare and Stallion team belonging to Tommie Flowers of North Carolina. One of the several breeds used as parentage for the modern American Belgian. Photo by Kristi Gilman-Miller

American Cream, a breed of Draft horse indigenous to North America and considered endangered because of low numbers of breeding stock. Photo by Carol Pshigoda.

bone, and sound lungs.

Several classes of equines have resulted and, depending on personal outlook, lines may be drawn and redrawn. This author draws them this way:

Heavy horses	Light horses
Draft ponies	Light ponies
Draft mules	Light mules
Mini mules	

As this book is meant to function as an operator's manual for the teamster's craft, I choose not to put a great deal of information here on breed specifics. We have included registries and associations in our resource directory and encourage the reader to contact them for in-depth particulars. For our purposes here, I'll mention several breeds of the two major classes, but this represents only a small portion of the many breeds found around the world.

Heavy horses

Clydesdale from the Valley of Clyde in Scotland
Percheron of North France
Suffolk of Suffolk County, England
American Cream, American Original
Shire of the east-central Shires of England
Ardennes of Belgium
Brabant of Belgium
Belgian (American) originally from Belgium
Boulanaise, France
Comtois, France
Noriker, Scandinavian
Spotted Draft, American

Draft Ponies

Norwegian Fjord
Haflinger, Austria
Icelandic

Light horses

Cleveland Bay of England
Standardbred of America
Morgan of America
Arabian
Thoroughbred of
England
Gelderlander – Germany
Holsteiner – Germany
Paso Fino
Appaloosa of America
Quarterhorse of America
Hackney of England

CLYDESDALES

(Right) This is perhaps the best-known draft horse breed in North America due to heavy beer company advertising. Even so, it is not anywhere near as popular, in terms of numbers, as other breeds. Perhaps this is because this breed is under-rated as a work-horse with many people complaining that height and bone are wrong for heavy work. Such is not the case, and this breed will play a larger part in the near future.

PERCHERONS

(Right) In numbers, this is the second popular draft horse breed with a solid history of contribution to farm and logging work. Certain criteria, used by some breeders who are looking for exaggerated action and artificial characteristics, may lead this stable old breed into genetic trouble in the near future. There simply aren't enough horses of the draft breeds to monkey around with the limited genetic pool and not pay the price.

SUFFOLKS

(Right) This breed has fewer numbers than even the Shire and is the least known. As the interest in good, honest hard-working farm and woods horses increases, this author believes that the SUFFOLK will gain in popularity. This breed has been selected over generations for work on the farm and has many favored characteristics.

SHIRES

(Top next page) This breed is scarce, with less that 500 registered purebred and percentage animals in 1980. There has been considerable recent importation from England and some constructive hope for increases in the breed in years to come. The genetic pool is small, so there is concern about the strength of the breed in generations to come, but the concern being there is good indication that the future is secure.

Don Anderson in the senior class at the Monroe Washington Draft Horse Extravaganza. John Erskine, the owner of Shires and wagon, rides along.

From the mid-seventies, a four abreast of Belgian horses at the Oregon State Fair.
The horses were owned and driven by the late Joe Van Dyke of Gaston, Oregon.

The late Ben Langston with the Quailhurst Clydesdale entry in a Men's Cart Class

Similar but different; Left, a Shire mare, below, a Clyde mare. While they share some important history and territory, these two British breeds continue to evolve distinctively.

BELGIANS

(Bottom page 50) Of the five major breeds of draft horses, the Belgians are the most numerous. The Belgian that has been developed over the years in North America has become unique enough genetically to be considered by some to be a breed of its own, separate from the origins. Certainly, to look today at the American Belgian and its distant parentage, is to see a world of difference.

Over the last 25 years there have been importations of what we erroneously refer to as the *European Belgians*. What we are more accurately referring to are the *Brabant*, the *Ardennes* and the *Comtois* breeds of France and Belgium. It is good to see the efforts to maintain, in North America, separate registries to help preserve the genetic purity of these breeds.

LeRoy Thoreson of Claremont, MN, with his mules at Heritage Farm's Iowa Plow Match, 1998. Photo by Bob Mischka. Draft mules and light mules have been a huge part of the construction and farming heritage of America. Though the mechanics of working mules is identical to working horses, many people will argue that the pysche of this hybrid animal requires special handling and more cleverness than is required to work horses. Similar arguments are made of the firm believers of the two primary political parties. Good mules are great partners.

Four Spotted Drafts pull a forecart and round baler during the 2000 Horse Progress Days held in Pennsylvania. This is a relatively new breed which is showing dramatic growth in numbers. Over the years the Lancaster, PA Amish community has favored Belgian mules with Belgian draft coming in second in numbers. Photo by Lynn Miller.

A Haflinger is hitched to a Mascot Sharpening designed and built 3 gang lawn mower displayed at the 2003 Horse Progress Days in Mt. Hope, Ohio. The Haflinger draft pony, originally from the alpine regions of Europe, has skyrocketed in popularity in the U.S. with Ohio being a true hot bed. These versatile, strong animals are well suited for all manner of uses. Their small size makes them particularly handy for small acreages and for people who might have difficulty throwing heavy harness up on a big draft horse.

Again at the 2003 Horse Progress Days two Haflingers demonstrate a KOTA disc harrow designed and built by B.W. Macknair company. Many Amish families have taken to the Haflinger and are enjoying brisk sales of foals.

Above: Rich Hotovy, a stalwart breeder and exhibitor of the Norwegian Fjord Horse, presents his four abreast at the HPDays Parade of Breeds in 2003. These unique scandanavian horses are multi-purpose and excel in harness. They have quiet dispositions and an attractive mid-size conformation. As with the Haflingers, the Fjords have enjoyed a meteoric rise in popularity all across North America.

Left: A handsome pair of Suffolks show off at the 2003 Parade of Breeds in Mt. Hope, Ohio. Endangered in 1980 and still threatened to some extent, the Suffolk has made great strides in recent years and the security of the breed would appear, at this time, to be headed in the right direction. We can only hope that, in North America, the 'utilitarian' criterias continue to rule breeding choices. The appropriate size, shorter cannon bone, and substantial heart girth are, in this author's estimation, more important than even the Chestnut coloring.

*Right: The **Ardennes** makes its home in France and Belgium. Genetically it is a first cousin to the **Brabant**. A third breed of the region, the **Comtois**, enjoys the same conformation in a sorrel color package. These three breeds formed the nucleus, over 100 years ago, for the **American Belgian**. As is obvious from the pictures, the **Brabant** and Ardennes, are unique breeds which warrant the recent work, in North America, to see them recognized and set apart from the **American Belgian**. These European breeds have seen their type heavily influenced by their secondary purpose as meat animals. Farmers have been making breeding choices which favor a frame that will gain weight rapidly. Though North American cultures, for the most part, frown on horse meat, it is favored in many parts of Europe and Asia.*

An Amish Standardbred buggy horse is hitched to a manure spreader outfit at Horse Progress Days proving once again that you can put any of the equines to work in harness, so why not choose the breed you are attracted to?
(Notice the relatively long cannon bone of this horse's front legs and remember that the muscle structure here is designed ideally for smooth extended gaits ideal for traveling down the road.)

(Below) Erik Stenvik of Norway with his unique hay rake hooked to a Noricker-style draft horse circa 1998. Harness designs, hitching systems and equipment vary widely throughout the world. Most of them have evolved to work with the farming cultures, terrain and animal types indigenous to the area.

American Cream

(See page 48) This endangered breed claims to be the only draft horse indigenous to this country. It's consistently unique coloring is a primary trait. A growing group of aficionados continue to work hard to bring back breeding stock numbers. Those folks looking to make a choice of work horse breed would do well to consider any of those animals which are endangered because of low numbers. Their populations will increase and the people who are on the ground floor of that growth will likely be the beneficiaries of respectable valuations when selling young stock.

The English Hackney demonstrating its high stepping gait.

An elegant Freisian horse shown at cart during the Parade of Breeds show at Horse Progress Days 2003.

This old etching typifies many of the European coach breeds.

Ponies of all sizes may be used in harness, from miniatures through Shetlands and up to the larger draft pony breeds. The team above were exhibited at the 2003 Horse Progress Days. Photo by Lynn Miller.

The late Allan Conder farmed with ponies. Here he is drilling seed with a four abreast of Shetlands. This drill would ordinarily be pulled by two draft horses.

This horsedrawn streetcar hauls people around Upper Canada Village, a living history museum in Ontario, Canada. This team is a lovely pair of old-style Morgans perfectly fitted to a new leather harness.

The Morgan horse is a true North American original with its entire strain said to have been descendants of one prepotent individual, the multi-talented multi-purpose stallion 'Figure'. The breed was named for his handler, Justin Morgan. The show ring has done far too much to over-refine the working traits out of the Morgan. In this day and age it takes a bit of traveling to find many of the "old draftier type" individuals. Photo by Kristi Gilman-Miller

Spotted Draft

(See page 52) Over this last quarter century, concerted efforts of several breeders have resulted in a strong start for a all new breed called Spotted Draft. Utilizing primarily Percheron and Belgian crosses to selected 'spotted' or paint stallions and then working to secure traits which would breed true, a solid foundation is being built. A breed to watch.

A photo taken in 1977 of Bob 'Bulldog' Fraser and his spotted team competing at Sandpoint, Idaho in the log skidding competition. 'Bulldog' has long been a lover of spotted horses. He trains horses and people (preferably women because he says they listen) up in the northwestern corner of Montana.

Draft Mules

Mammoth Jacks are bred to draft horses resulting in draft mules. Far and away the most popular cross is with the Belgian breed. Due to the heat tolerance and extreme regional popularity Mules are certain to always enjoy a place in the South. In the carriage trade of New Orleans, city regulations allow that mules may work in warmer temperatures than may horses.

A Belgian draft mule at work in Waitsburg, Washington. Photo by Lynn Miller.

A championship draft mule from the 1977 Draft Horse Extravaganza in Monroe, Washington. Photo by Lynn Miller.

Twelve head of red sorrel Quarter horse cross mules hitched to a three bottom trail plow, Jay Thomas driving in 1979. Jay and his father, the late Don Thomas, put together 20 of these mules to use annually at their Waitsburg, Washington farm to recreate the big combine hitches. They also traveled around the northwest showing their hitch at many draft horse shows.

The Blethen family of Washington state maintained a Suffolk hitch. This was their team entry in the lady's drive at the 1980 Draft Horse Festival in Eugene, Oregon.

Below: Bobbie McGhee shows her outstanding sorrell Percheron stud colt to a championship. This youngster has the scale and conformation to grow into a tall hitchy type.

Below: Donna Anderson, in 1980 at the Draft Horse Festival, guides her American Shire team through a tight log skid pattern. Donna's the daughter of legendary teamster and Shireman Mel Anderson. And her brother Bill is a champion teamster and harness maker.

For many Shire breeders there are two very distinct types within the breed; the English type and the American type. The American Shire is said to be shorter coupled and lower to the ground,

Brabant horses at work on a seed drill in Belgium. Photo by Matilda Essig.

Don Lee, an Oregon horselogger, with his old-style Percheron team in the log skid.

Mitchell's Contributor Joe of Iowa, an outstanding American Belgian stallion of the old-style. Photo taken in 1979 by Lynn Miller.

The author's American Belgian stallion, Abe. Photo by Kristi Gilman-Miller.

ANATOMY

The illustrations best show the position of the various parts of the horse anatomy. (Some discussion of form and function was gone through in Chapter Three.) I might suggest that the most important parts of the horse, in order to priority, include:

1. Feet (up to and including fetlock joint.)
2. Hocks
3. Knees
4. Legs (Note: if these first four are not in order, there is no need to look further. Look at each new horse from the ground up.)
5. Eyes
6. Teeth
7. Shoulder
8. And so forth (from here on all else is equal.)

There are volumes written about the anatomy of the horse and this text is not the place to go into the subject in any depth. There are some important things to mention, however.

THE BOTTOM

When a horseman refers to the "Bottom" he is usually speaking of the anatomy of a horse from the knees and hocks down. Whatever the intended use, if it does not have a good sound "bottom" (or feet and legs), a horse is of no value. The drawings in this text illustrate the individual parts of the bottom. If the reader has a particular concern, interest or problem regarding something very specific to this or any part of the horse's anatomy, the author recommends that you seek out specialized sources of information.

The bottom starts with the feet, the first place you should look at any new horse. A close examination of any horse's foot discloses a unique structure and one which has suffered a wide variety of man's theories through the ages. For a horse's hoof to be sound it should demonstrate a strong, yet slightly pliable hoof wall without excessive long vertical cracks. The animal should stand naturally with a hoof wall angle which matches the pastern angle, which in turn should match the shoulder's angle (see page 34, *shoulder angle*). The coronary (coronet) band should be clean and smooth without obvious wounds or, when examined by touch, without calcium-like growth under the skin, just above the coronary band. Examination of the underside of the foot should find a healthy frog and heel with no sign of black infection alongside the frog in the quarters (thrush). Scraping of the sole should not discover

any obvious stone bruises (blood marks). On the whole, the foot should present a balanced look, (see page 65).

From the coronary band to the fetlock runs the pastern. This assembly from fetlock through pastern to the foot is the most critical section of the entire horse. Without free and correct motion of these parts a horse cannot be expected to do any work whatsoever. It would not then matter how beautiful the topside of the animal was.

From the fetlock joint up to the knee runs the cannon bone. The proportion of the cannon bone has a direct relationship to the efficiency of the work horse. (See chapter three MECHANICS OF DRAFT.) The knees of the front legs serve to balance the animal while the hock joints of the back legs must work under greater strain as the legs move the mass in a digging-like motion. The hock joint is of critical importance to the ability of the horse to pull heavy loads. A good, clean, strong hock joint which exhibits straight-ahead motion is desired. The hock joints should not be set too close together nor too far apart but rather in a symmetrical balance. Pages 63 & 67 illustrate the relationship and proportion of some of the bone structure of the "bottom."

THE TOP

There is a great deal of disagreement in the horse world about what constitutes the ideal "top" for horses of given breeds or types. Long backs vs. short backs; long necks vs. short necks; these are a few of the arguments you might hear. First of all, there is lots of opportunity for optical illusion in the horse's form, which is to say that a long-legged horse might give the illusion of being longer in the back or a long-backed horse might give the illusion of being shorter-legged. It takes a skilled eye to see the true form of the horse for what it actually is. And for the person interested in form as it relates to capacity for work, the author's opinion is that personal experience coupled with some sprinkling of personal aesthetic preference is the best guide to making that determination. The preceding chapter gives some specific information explaining this form and function question regarding the work horse.

The work horse should exhibit ample heart girth which suggests good lung capacity. The horse that is expected to work in collar and harness must have a correct and healthy shoulder both from the inside and outside. Page 34 illustrates the relationship of the bone structure to how the collar sits and suggests how

anatomy

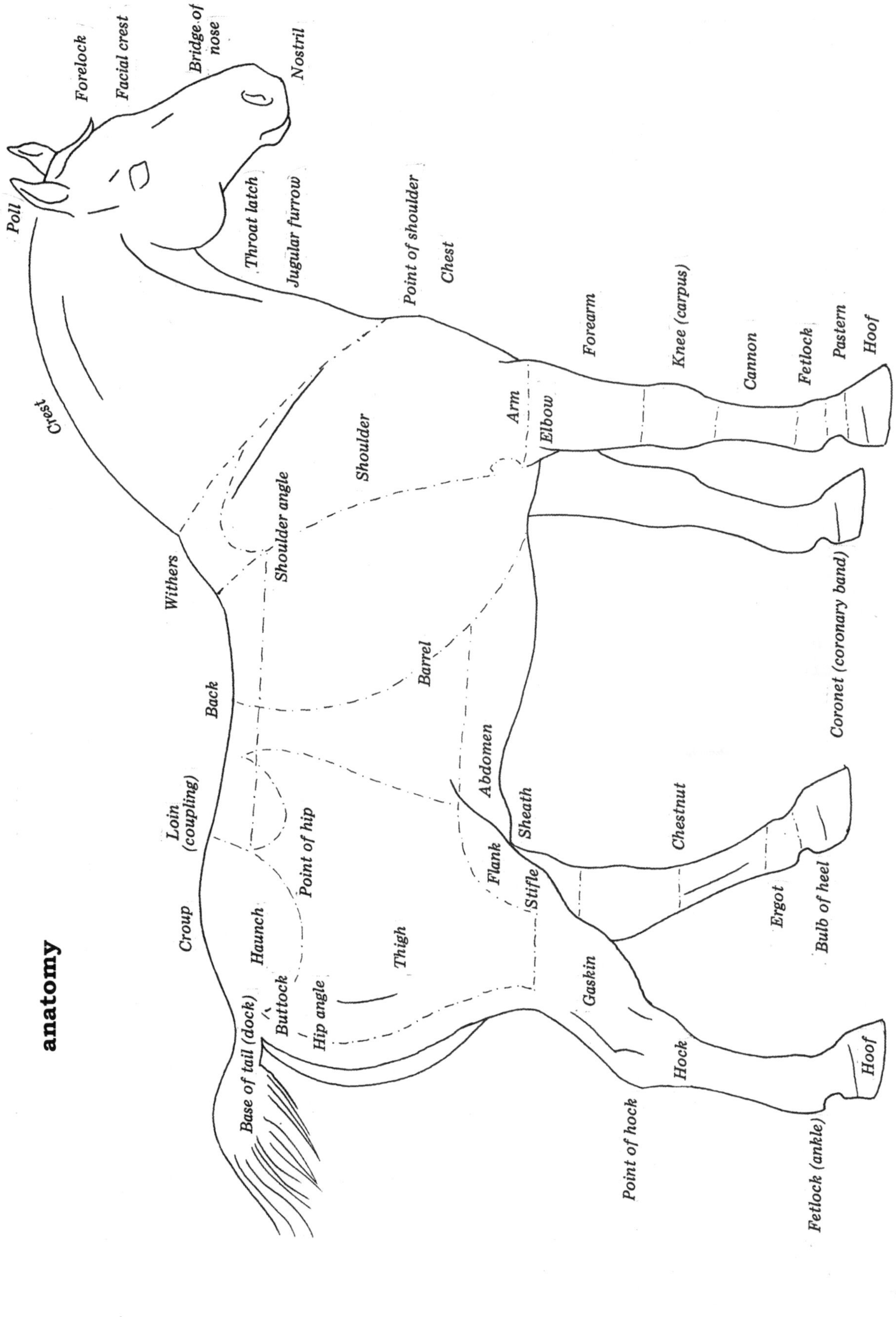

Poll

Forelock

Facial crest

Bridge of nose

Nostril

Throat latch

Jugular furrow

Point of shoulder

Chest

Forearm

Knee (carpus)

Cannon

Fetlock

Pastern

Hoof

Crest

Shoulder angle

Shoulder

Arm

Elbow

Withers

Barrel

Abdomen

Sheath

Chestnut

Coronet (coronary band)

Back

Loin (coupling)

Point of hip

Flank

Stifle

Ergot

Bulb of heel

Croup

Haunch

Buttock

Hip angle

Thigh

Gaskin

Hock

Point of hock

Fetlock (ankle)

Hoof

Base of tail (dock)

skeletal structure

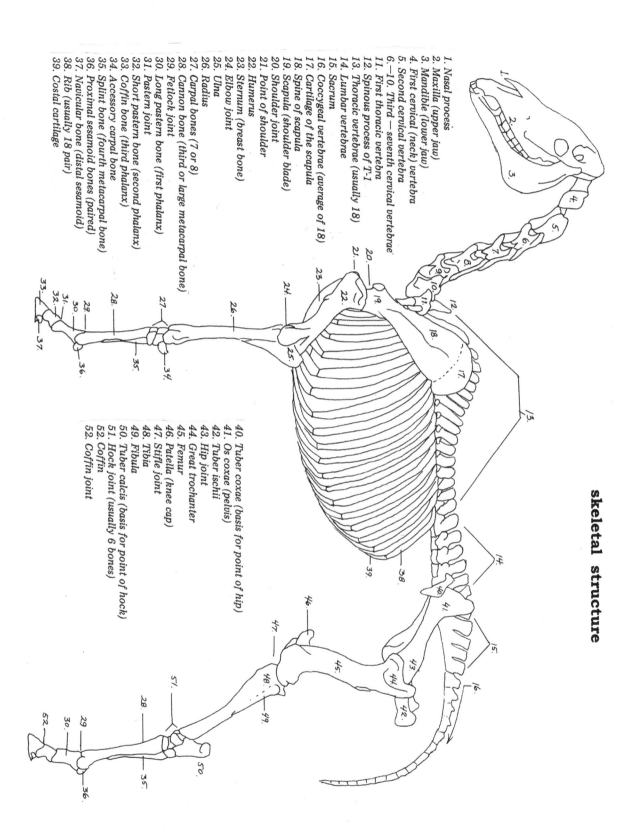

1. Nasal process
2. Maxilla (upper jaw)
3. Mandible (lower jaw)
4. First cervical (neck) vertebra
5. Second cervical vertebra
6.—10. Third — seventh cervical vertebrae
11. First thoracic vertebra
12. Spinous process of T-1
13. Thoracic vertebrae (usually 18)
14. Lumbar vertebrae
15. Sacrum
16. Coccygeal vertebrae (average of 18)
17. Cartilage of the scapula
18. Spine of scapula
19. Scapula (shoulder blade)
20. Shoulder joint
21. Point of shoulder
22. Humerus
23. Sternum (breast bone)
24. Elbow joint
25. Ulna
26. Radius
27. Carpal bones (7 or 8)
28. Cannon bone (third or large metacarpal bone)
29. Fetlock joint
30. Long pastern bone (first phalanx)
31. Pastern joint
32. Short pastern bone (second phalanx)
33. Coffin bone (third phalanx)
34. Accessory carpal bone
35. Splint bone (fourth metacarpal bone)
36. Proximal sesamoid bones (paired)
37. Navicular bone (distal sesamoid)
38. Rib (usually 18 pair)
39. Costal cartilage

40. Tuber coxae (basis for point of hip)
41. Os coxae (pelvis)
42. Tuber ischii
43. Hip joint
44. Great trochanter
45. Femur
46. Patella (knee cap)
47. Stifle joint
48. Tibia
49. Fibula
50. Tuber calcis (basis for point of hock)
51. Hock joint (usually 6 bones)
52. Coffin
52. Coffin joint

LOCATION OF COMMON DEFECTS

1. Poll evil
2. Defective eyes
3. Defective teeth
4. Roaring
5. Sweeney
6. Sprung knee
7. Ring bone
8. Bad hoof
9. Side bone
10. Splint
11. Bog spavin
12. Capped hock
13. Curb
14. Bone spavin
15. Fistulous
    withers

Top
Bottom

MUSCLE STRUCTURE

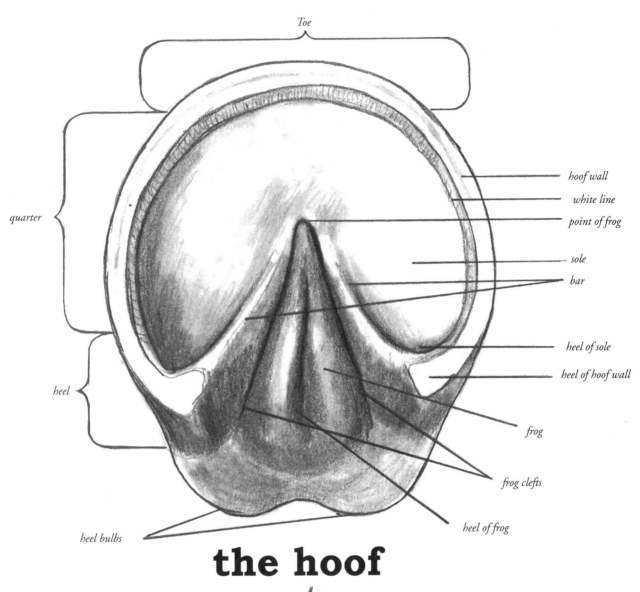

Toe

hoof wall

white line

point of frog

sole

bar

quarter

heel of sole

heel of hoof wall

frog

heel

frog clefts

heel of frog

heel bulbs

the hoof

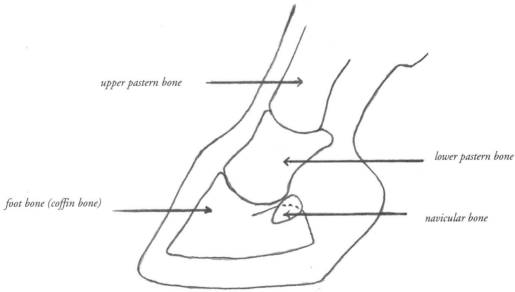

upper pastern bone

lower pastern bone

foot bone (coffin bone)

navicular bone

ANGLE OF THE PASTERN

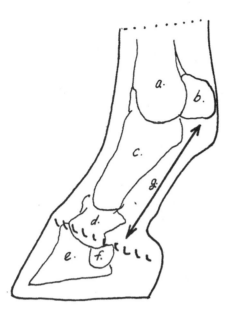

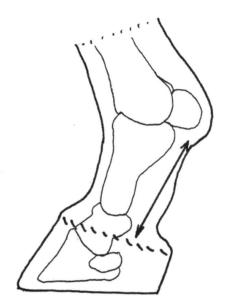

a.) Illustrating the correct angle of the pastern and slope of the hoof wall. The line from b. to f. (marked g) corresponds to the position of the navicular tendon which must expand and contract with movement.

b.) Showing the horse at the first stage of forward motion with an extended, or stretched, navicular tendon.

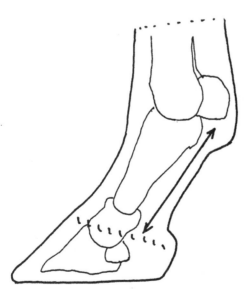

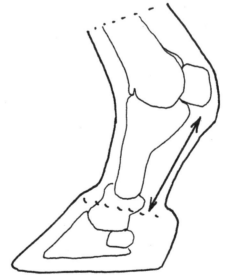

c.) Illustrating an incorrect angle caused by too long a toe and too short a heel. The angle causes a flexing of the tendon.

d.) Showing the incorrect foot at the first stage of forward motion with added tension on the naviular tendon. This causes pain, resulting in a quick 'snappy' action and eventual navicular lameness.

important it is that there be no abnormalities in bone, muscle or skin.

The horse's head will tell a great deal about the animal and should be regarded carefully when purchasing a new horse. Ears and eyes tell you of the animal's more obvious personality traits such as gentleness, meanness, docility, or nervousness. Look for a calm, quiet, yet alert eye and ears which are comfortably at attention. Teeth can, to the experienced or initiated, determine the age of the horse as there are distinct changes which occur each year. The nose can be an early warning of respiratory infection, which can be chronic or serious in horses and is most often highly contagious. The general balance of conformation of the head can be an indicator of capacities or character. The most commonly considered relationship are the placement of the eyes, whether too close, too high, too low, etc. Generally look for what appears, to the intelligent view, to be a balanced head.

As explained in the FEEDING chapter, the horse has a unique and highly sensitive digestive system. The discussion in this chapter on DISEASES AND DISORDERS touches on the subject of the various causes of disorders of the digestive system.

FUNDAMENTALS OF CARE
unsoundness

All of the humorous stories you may have heard relative to horse-trading have one serious implication in common; it is that the selection of a sound and desirable horse is often difficult. The best way to determine soundness is with a careful systematic examination of the horse.

Lameness is not always readily apparent, so it is important to have the animal walk and run while you observe. This should be done more than once, having someone turn the animal to the right and left while trotting.

The defects map on page 64 should be used to aid in a careful examination for these possible problems:

Poll-evil. Found at the top of the head, or the poll, and caused from a bruise, indicated from inflammation or a scar. Expect that it will recur and cause problems especially with halter and bridle wear.

Fistula. Found at the withers and caused by a bruise, similar in some respects to poll-evil, but should not be confused with collar sores; indicated by inflammation or scars.

Ring-bone. Bony enlargement found at the front

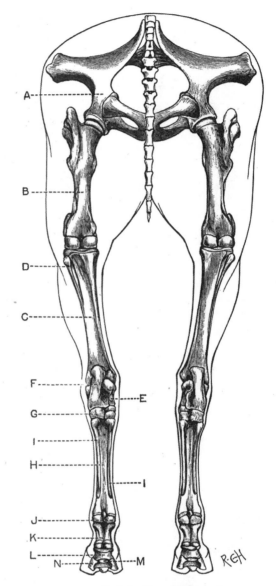

–Posterior View of Pelvis and
Hind Limbs of Horse

A, Pelvis. B, Femur or thigh-bone. C, Tibia or lower thigh-bone. D, Fibula. E, Astragalus. F, Calcaneus, forming point of hock. G, Cuboides. H, Large metatarsal or canon-bone. I, I, Small metatarsal or splint-bones. J, Sesamoid bones. K, Os suffraginis or large pastern. L, Os coronæ or small pastern. M, Navicular bone. N, Os pedis or Foot bone.

and sides of the pastern, which causes lameness when developed to sufficient size to interfere with the action of the joints and tendons.

Side-bone. Side-bones can best be seen from the front and occur on the sides of the coronary (coronet) band (top of the hoof); on the forefeet they interfere with action and may cause lameness. When not able to detect them by sight, they may be found by slight

Narrow Breast

Broad Breast

pressure with the fingers. If the coronary band yields to pressure it is healthy; hardness is an indication of unsoundness.

Splints. Bony prominences found on the inside of the cannon, just below the knee, may cause lameness when first developing, or when close enough to the knee to interfere with movement. They often disappear in young horses.

Bog Spavin. An enlargement of the natural depression on the inner and front part of the hock, which is soft to the touch.

Bone Spavin. Bony enlargement on the inside of the hock where the thick part tapers into the cannon. Pick up the foot, hold flexed, release and start the horse to trot. Lameness indicates bone spavin.

A **blemish** is a defect which detracts from the appearance but

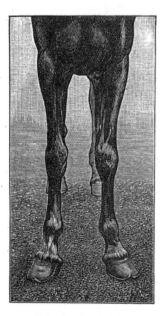

Toes turned out

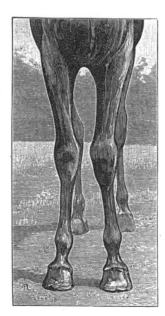

In-kneed

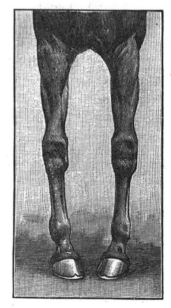

Toes turned in

front legs: telling the good from the poor

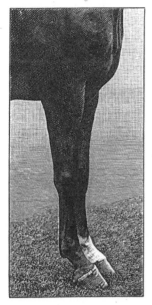

Good Pasterns

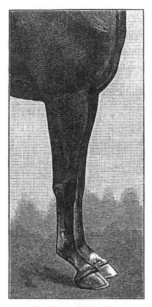

Long sloping Pasterns

Short sloping Pasterns

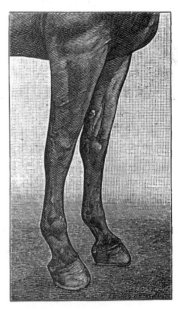

Calf knees

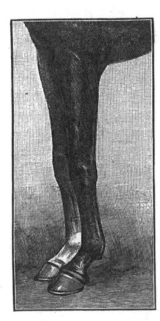

Bowed knees

illustrations from The Horse *by Axe 1902*

Good fore-arm and cannon

Weak fore-arm and cannon

front legs: telling the good from the poor

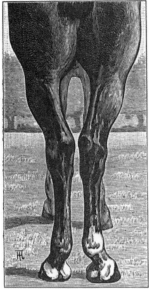 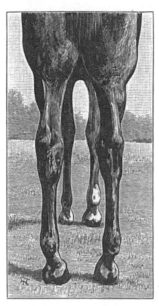

Cow-Hocks   Bow-Legs

hind legs

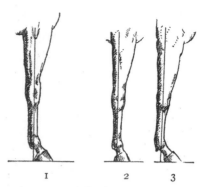

slopes of the pastern

The above three illustrations demonstrate strong and poor front leg alignment as related to angle of the pastern. Number one shows balance and strength. Numbers two and three show weakness and discomfort.

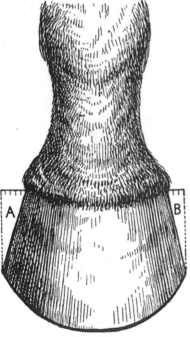

Normal Foot: front view, showing slopes of (A) outer wall and (B) inner wall

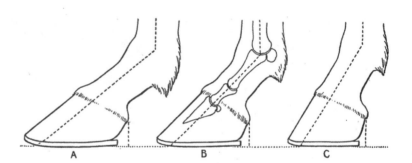

Well-proportioned and Ill-proportioned Feet

A, Foot too long and heel too low.   B, Well-shaped foot.   C, Heel too high.

Shoe Fitted Short at the Heel

hoof angles and shoe fit

An "Eased" Heel

Shoe with Level Bearing

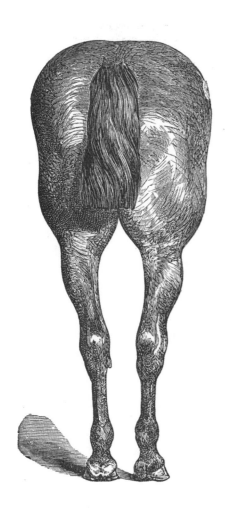

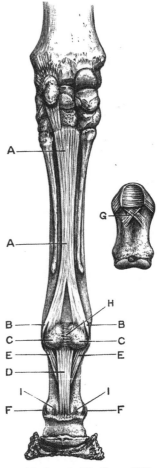

good front and back

Attachment of the Sesamoid Bones
to the Skeleton of the Leg

A A, Suspensory ligament. B, B, Outer and
inner branches of the same. C, C, Outer and
inner sesamoid bones. D, Superficial sesamoid ligament. E, Deep sesamoid ligament.
F, F, Lateral phalangeal ligaments. G, Crucial
sesamoid ligament. H, Intersesamoid ligament. I, I, Posterior phalangeal ligaments.

clean head, kind eye

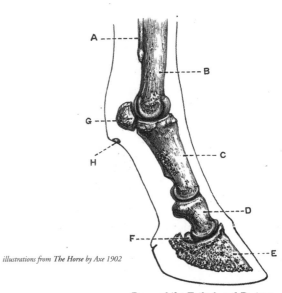

illustrations from The Horse *by Axe 1902*

Bones of the Fetlock and Pastern

A, Splint-bone. B, Canon-bone. C, Suffraginis or
first phalanx. D, Second phalanx. E, Pedal bone.
F, Navicular bone. G, Sesamoid bone. H, Ergot.

not the usefulness of the horse.

diseases and disorders

This does not pretend to be a detailed veterinary discussion and should not be depended upon as a final aid in diagnosis or treatment. The teamster should look elsewhere for help with diseases and disorders of the horse. The information presented here is a beginning to hopefully aid the teamster in understanding the reasons for the best care of the horse.

The horse is the healthiest of all farm animals when properly managed. Troubles begin when horses are closely confined, overworked, improperly fed or watered, or under-exercised. Under these conditions, the horse is subject to numerous ailments. The ailment may be a discomfort which readily responds to correction of the cause. It may be a disease of nonspecific type, such as colic (acute indigestion), which is caused by a sudden change in ration, feeding frozen or moldy feed and/or other physical intrusions on the digestive system. Or, due to lowered resistance resulting from improper management, the horse may have a specific infectious disease which may be transmitted to the other work stock on the farm if not properly managed.

Recognizing Abnormal Conditions. The importance of recognizing any abnormal condition in the early stages cannot be over-emphasized. Anyone who cares for the stock regularly can recognize the first symptoms of discomfort or disease by the change in the general appearance, behavior, and bearing of an animal. A close checkup should be made at once in regard to the feed the animal has had, the conditions under which it has been kept and whether the animal has been standing idle or possibly is overworked.

Every factor in the immediate history of the case should be considered in order to give the horse proper care. The attendant should put the horse in a box stall where it can be made as comfortable as possible. Notice should be taken regarding respiration, appearance of the coat and skin, presence of fever, and rate of pulse. All secretions and excretions, as well as further behavior, and evidence of pain, swelling, tenderness, and loss of function should be noted.

In many cases good care and correction of the feeding practices are all that are necessary. In other cases a simple laxative, such as a quart of mineral oil, is all that is needed. In still others, rest may be the cure. The first thought, however, should be to look for the cause of the ailment and remove it.

In what may appear to you as an emergency situation, time and the useful life of your animal may be saved by you being able to give good information in the initial phone call to a veterinarian.

Here's a fairly basic checklist for evaluating physical status:

General health observations

a. General condition- weight, hair coat, muscle tone. (i.e. thin, rough hair coat, poor muscle tone, etc.)

b. Attitude, behavior. (i.e. cranky, sullen, depressed, disoriented, anxiously circling, repeated pointing to abdominal cavity, getting up and laying down, etc.)

c. Exercise tolerance (i.e. tires easily)

d. Appetite (i.e. always hungry versus refuses to eat)

e. Posture, gait changes (i.e. head down, favors leg, legs close in under, legs spread or splayed, ears drooping. etc.)

f. Cough, Sneeze, Abnormal breathing sounds or effort.

g. Nasal discharge

h. Feces, urine (i.e. discoloration, loose, obvious pain in execution, etc.)

Objective Data

a. Rectal temperature

b. Heart rate, Pulse rate (strength and rhythm)

c. Respiratory rate (depth and pattern)

d. Mucus membrane color, capillary refill time, moisture.

e. Hydration Status

f. Jugular refill

g. Gastro intestinal motility

h. Digital pulse

Many of the above items will be completely foreign to the uninitiated. We suggest that a copy of these two lists be shown to your veterinarian with the request that he or she demonstrate, or tell you, how to arrive at these measurements. In this way you will be prepared, should the remote possibility of an emergency occur, to phone in an initial set of observations. These then MAY allow the veterinarian to suggest an immediate course of action, and/or prepare themselves to deal with the situation.

Here are the normal values for an average horse at rest:

Rectal temperature 99 to 100.5 degrees F
Pulse Rate: 25 to 60 beats per minute.
    rhythm: regular
    strength: strong

Respiratory rate: 8 to 20 breaths per minute
   depth: moderate
   effort: none apparent
Capillary refill rate: 1 to 2 seconds
   Mucus membranes: pink, moist

CARING FOR THE SICK HORSE

Here's a list of items which we suggest you keep on hand for any emergency (check with your vet to see how they would modify or add to this list).

   a veterinary recommended equine Systemic
      Antibiotic with suitable needles and syringes
   Banamine paste (muscle relaxant)
   Butazolidin (pain killer)
   Aspirin (pain killer)
   Ichthammol Ointment
   Nitrofurazone Ointment
   Betadine
   Large wound dressing pads
   Wound wrap
   Duct tape

Regardless of the ailment, the care the horse receives will greatly affect the progress of its recovery. Sick horses are usually nervous and should be kept as quiet as possible. They should always be handled with the greatest patience. The stalls should be comfortable, free from drafts, and the bedding kept fresh and clean. The feed should be laxative and easily digested. A little fresh grass or apples may be given if the horse's appetite is poor. Grooming with a soft brush or cloth, and frequent rubbing of the legs, will aid the circulation and help to throw off the body waste. Access to a quiet paddock for free exercise is beneficial but the ill horse should not have to compete with any other animals.

If there is any chance that a disease may be infectious, the animal should be kept isolated from all other susceptible individuals. The well animals may be protected by keeping the sick one in a box stall or paddock that does not connect with any other stalls. The sick horse or mule should be fed and watered from individual buckets. These utensils should be thoroughly sterilized before being used by well animals again.

If there is any evidence that the animal is seriously ill or badly injured, waste no time in securing the advice and services of a licensed veterinarian. Stay away from self-styled horse doctors but in the same vein; do yourself and your animal the service of checking up on the reputation and skill of any vet you might choose. A diploma and a license do not, unfortunately, guarantee that the vet holding them is competent and honest.

Digestive disorders. Digestive ailments, most falling under the catchall name of "colic" are the most common ailments of work horses and mules. These ailments may occur either in mild or acute form and are caused by indigestible feed, spoiled feed, improper feeding methods, improper watering, or by any other factor that might disturb, block or disrupt the digestive system.

During **colic** attacks the animal refuses to eat his feed, lies down, gets up, paws the ground and shows evidence of distress. The horse will probably make circling motions apparently pointing with its head to the gut. In colic, the pain may be sudden and intermittent, the animal appearing to be relieved between attacks. The respiration and pulse increase decidedly and the animal may perspire profusely.

If the condition appears at all severe, fill a quart size soda bottle (with a long neck and narrow opening) with mineral oil and place in the corner of the animal's mouth. Hold the chin up with the head just slightly up from level and pour the oil slowly down the throat. The object is to get most if not all of the oil down the animal to work as a laxative and lubricant.

Call a veterinarian immediately!

Ask the vet if you should administer a dose of the muscle relaxant *Banamine*. While waiting for the vet, in a severe case, lead the animal around, in a paddock-like area, at a slow steady walk, discouraging the animal from lying down.

Colic can range from a mild upset stomach that corrects itself, all the way to a twisted or telescoped intestine which usually results in death. Do not treat any symptom of colic lightly. Make an intelligent, deliberate attempt to determine the exact cause of the colic but realize that in some cases you may not be able to find the cause. The very best insurance against colic is prevention through good care.

Exercise related Myopathy or Azoturia. This is a condition which goes by many names including: azoturia, exertional myopathy, myositis, monday morning disease, overstraining disease, and paralytic myoglobinuria. It is not the same as *tying up syndrome*. Azoturia, as we will call it, is caused by over-feeding and under-exercising. It comes on quickly after the hard working horse on full ration has been standing idle for a few days still on full ration. It is believed that this condition is caused by a rapid build up of lactic acid in the horse's system. The blood stream can't get rid of the excess and the acid builds up in the muscle cells resulting in varying degrees of muscle damage. After going a short distance, the horse appears confused, perspires freely, trembles, and shows difficulty in controlling the hind parts. Other signs may cause the lay person to confuse this condition with colic.

If a working horse shows these symptoms, unharness immediately. Do NOT move the animal. Apply warm compresses and muscle massage. Give 2 tablespoons of Baking Soda orally at first signs. Administer Butazolidin or Banamine. Warm blankets should be placed over the loins. Rest. After recovery, daily exercise with reduced feed when not working should prevent recurrence. Prevention comes with a careful change in ration during rest periods. (See chapter five).

Tying Up Syndrome. This is the look-alike disease to *Azoturia*. One major characteristic is that this disease follows prolonged exhaustive physical activity. It is the result of muscle energy depletion. First aid therapy is exactly the same, but follow-up treatment is completely different. Minimize total confinement. Feed more hay and less grain or avoid grain when the horse is not used. Practise slow warmups and slow warmdowns. Greater care in conditioning. Provide water and electrolyte supplementation during prolonged exercise.

Influenza is an infection of the mucous membranes in the respiratory tract. The horse seems depressed and stands with its head down. It has a high fever and loses its appetite. It may sneeze and cough, and the mucous membranes of the eyes are often inflamed.

Consult a vet who will advise as to which antibiotics, if any, are to be administered. Keep the animal to itself in quiet, clean, well ventilated, but draft-free quarters and on a light laxative diet.

Distemper is an infectious disease which may be contracted directly from animals suffering the same or from watering troughs or stabling used by infected animals. Young animals are most susceptible.

The ailment affects the air passages and the glands in the throat. The disease is characterized by loss of appetite, fever, nasal secretions, formation of pus in the air passages, and later by abscesses under the jaw. The abscesses break after a few days and the pain, which has been severe, is relieved.

The attention or recommendations of a competent veterinarian plus rest, comfortable quarters, and good care are the necessary requirements in average cases. Isolation to prevent spread of the disease is important.

Many of the diseases of work horses or mules have symptoms in the early stages that are so much alike that it is impossible for an inexperienced person to distinguish one from another. Any disease of a more serious nature than those mentioned above, as well as severe cases of the same, should be tended to by a competent veterinarian.

Injuries. There are three types of injuries commonly found with work horses; first, lacerations, such as wire cuts; second, bruises, such as kicks from other animals, and third, sores and wounds caused by ill-fitting harness.

Lacerations (Bad Cuts). Excessive bleeding must be stopped. Use of a commercial astringent, or shreds of sterilized cotton packed lightly into the wound may hasten clotting of the blood. When the bleeding stops, any foreign matter that has not been washed out by the blood flow should be carefully washed out with a disinfectant solution while washing the entire surrounding area. If the skin is badly torn, stitches may be necessary. If so, a veterinarian should make the determination. Unnecessary stitching could cause unwanted infection. There are a variety of healing powders and ointments available on the market which may help to speed up healing.

Subsequent care of the injured animal is important to prevent complications. If the animal runs through dewy grass or weeds, pollen and other irritating substances may get into the wound causing inflammation, preventing speedy healing. This sort of inflamed, irritated, open wound is a prime target for the development of "proud flesh" or abnormal tumor-like growths. Bandaging may be necessary to prevent foreign matter from getting into the open wound. A clean, dry stall or barn lot would be preferred housing if the wound is in a place that is highly susceptible to dirt or foreign irritants.

Normal healing often causes itching, and in order to relieve this, the animal may lick or bite the wound. In such cases the injury must be protected. Sometimes bandages work, sometimes they don't. Confining the movements of the horse's head by a short halter rope fastened, is one possible course of action.

Bruises are similar to lacerated injuries, but the skin is not broken. If the bruise is sufficient to cause considerable internal bleeding, death may result, or other less final, but nonetheless dangerous, results. If the bruise is not severe, hot compresses applied to the part for two hours, followed by cold compresses for an hour or two, may give relief. A liniment rubbed into the bruised area will help. Absorbine and Bigeloil are two which work very well. The author's experience with bruises have resulted in a preferred course of action. This plan is particularly useful if the bruised part is an active muscle. Through a veterinarian's office secure some "DMSO" and rub this substance on the affected area. Wait one hour and wash or hose down the bruise with cool tap water. Repeat this three times a day until swelling and/or discomfort subsides.

Harness injuries occur frequently. They are most always the result of ill-fitting or poorly constructed harness or because of improper methods of breaking in or toughening the shoulders for work in the collar. The pressure causing the injury should be removed. The injury should be treated as either a lacerated wound or a bruise, whichever the case may be. In mild cases the affected area may be washed with a cold salt-water solution and dried. There are several gall cures on the market that are good to use in cases of chronic irritations from ill-fitting harness. The author has had tremendous success in heading off sore shoulders and harness sores with liberal applications of *Ichthammol* ointment.

Collar sores. After removing the harness from working animals this author always checks the shoulders for hot spots, soft spots, or raised spots. After careful cleaning, a thick application of *Ichthammol* is applied. It works as a drawing agent and, in most cases if applied early enough, will prevent collar sores.

Parasites. Work horses are susceptible to many parasites both external and internal. Parasites reduce the vitality and efficiency of work horses and in some cases can be directly linked to permanent damage and even death.

Internal Parasites. Among the common internal parasites that infest horses are the bots, stomach worms, intestinal round worms, pinworms and tapeworms. These and other internal parasites cause indigestion and colic affecting the walls of the stomach and intestines and indirectly affecting other organs and ultimately the entire condition of the animal.

The manifestation of the presence of parasites in work stock is often so gradual that the owner may not suspect it. The animals which he considers "hard to keep," or requiring more feed and special attention, are usually the ones infested with the parasites. Whenever the condition of the horse indicates unthriftiness without the presence of fever – unless the cause can be attributed to improper feeding or management or

abusive attitude – it is very probably due to the presence of internal parasites.

There have been frequent changes, in recent years, in popular programs for the control or eradication of internal parasites. Research continues as there are many unanswered questions about the real effects of the parasites and of the best methods for control. The most recent research has indicated that complete eradication of internal parasites is now possible with new high-powered chemistry and careful timing. However, further research indicates that to rid the animal of all parasites may be guaranteeing disaster as the presence of small numbers of worms in the system seems to generate natural antibody action that is not only beneficial but possibly essential. For without these antibodies, an animal that is infested with a new crop of parasites would suffer death.

The very best system for parasite control must be carefully tailored to the individual animals. By taking fresh stool samples from each animal and having them examined every 6 to 8 weeks by a veterinarian the exact nature and extent of infestation will be known and the correct medicine can be prescribed. This also keeps you in tune with the veterinarian's bank of information from changing research. To simply purchase worming medicine at a local feed store and follow the "company's" recommendations for dosage and frequency is to invite varying degrees of failure. By tailoring the program to the individual animal's actual infestation you guarantee that your efforts and money will be well spent.

Medical treatment of internal parasites should be supplemented by programs for sanitation control. Stables should be cleaned regularly and manure piles should not be available for contact with the horses. Stock that is known to be heavily infested should be kept clear of other animals and care should be taken not to use the same feed and watering areas and pastures. Rotation of pastures should be an important program for the farm.

External Parasites. There are many external parasites that are a source of worry and trouble to work horses. They irritate the skin, causing various types of skin diseases. Like internal parasites, they cause a general unthriftiness in the animals.

Among the most common external parasites are lice, mites, fleas, ticks, and flies of many kinds. The discomfort of the animal is readily noticed, and usually the course of misery is easily diagnosed. There are two approaches to treatment. One is to keep the pests from the stock. The other is to get rid of the pests altogether. There are a variety of chemical and organic treatments on the market which work in varying degrees to keep the pests off the animals. The author has used a couple of different brands with some success. However, the preferred course of action is to employ a variety of natural and chemical means to control the population of unwanted pests. There are firms that sell natural

predators of flies and other insects. In other words, insects which eat the unwanted insects. There are also mechanical means with new traps and electric "bug zappers" gaining in popularity. The best program will be the one which is designed specifically for the farm in question using the best means, after careful consideration of the consequences and the fragile natural balances that exist.

GROOMING

In order that horses may work most effectively, and with the greatest comfort, attention must be given to grooming. Lack of adequate grooming impairs the health and efficiency of work horses. Grooming improves the general appearance of a horse, and, what is more important, removes the loose dirt and internal waste of the body, which has been exuded through the pores of the skin. If this waste is not removed by thorough grooming, the pores of the skin get stopped up, normal bodily functions are impaired and the general health of the horse is hindered. Poor condition is indicated by a dull, harsh skin and a tough, dry appearance of the hair. The skin of a well-groomed, healthy animal is pliable, and soft, with the hair glossy. There are those who do not believe it, but regular grooming has been proven to improve the action of digestive organs and the utilization of feed.

The amount of grooming necessary will vary with whether horses are pastured and out-of-doors or if working and housed. Horses that are pastured do not require much grooming. Under these conditions there is less perspiration, and the waste products are more generally thrown off through the bowels and kidneys. Fast or active work, together with heavy feeding and a degree of confined stabling causes free perspiration, which throws waste through the skin and makes regular grooming necessary.

Care and thoroughness are essential in grooming. It is a good plan to groom the horses at night, for they rest better afterward. A good, brisk brushing will be sufficient the next morning before going to the field. Turning the horses out at night, or allowing them to roll, only partially removes the need of the evening grooming. Only the high spots are touched in rolling, and very little of the body waste material is removed in

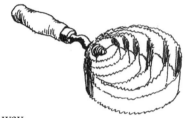

this way.

Grooming is essential for teams that are at hard work. Special care should be taken to remove all loose dirt and hair from the shoulder and other parts where the harness works. Care must be taken, however, not to use a hard curry comb over the shoulder directly before or after work as it could cause soreness or irritation that would result in a gall. Animals that have been wintered largely on roughage and then changed to a ration with grain will respond quickly to regular grooming. Farm horses are not usually groomed when they are idle. During the winter they grow a heavy coat. Clipping off this coat in the spring can make grooming easier and prevent excessive sweating, but great care should be taken not to let the horse get chilled. They will have to be protected from the cold for a time. If the teamster prefers not to clip, sweat scrapers with serrated edges are available which, if used in the grooming process, aid greatly in the removal of shedding hair and speed the animal to their seasonal coat.

CARE OF THE FEET

For an appreciation of the shape of the horse's hoof see page 65. The feet of horses and mules usually remain healthy and the walls tough when at work. The feet should be trimmed and kept square and plumb at all times or else excessive strain on the tendons may result. Unless correcting a temporary or permanent defect, it is not necessary to shoe the farm horse as most work is done on dirt with little concussion. For highway work, winter work or logging, it may be necessary to have shoes kept on the animals.

Care of feet is one of the areas of horse management where man through the ages has done his own

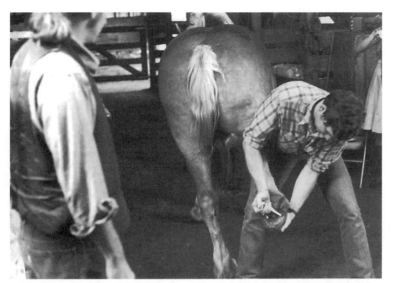

Learning to be able to pick up your horse's feet and clean them is an important skill for the teamster. Your horse's training and your safety should always be considered.

ego some good and the horse a great deal of harm. The so-called experts do NOT agree to this day what constitutes the best care. When you consider how critical the horse's foot is to movement and work you realize that it is impossible to either cover the subject in adequate depth in this text or to give a beginner the tools necessary to understand, let alone work on the foot. However, if a person chooses to depend on work horses for a measure of power then he or she will be in line for some schooling in the farrier's art. The cost of having a farrier come regularly and trim and/or shoe your animals will quickly become prohibitive to most working budgets. Add to this the fact that you who work with the animal all the time are better equipped mentally to do the hoof work. But that will come with time and patience. In the beginning you will need to depend on a competent farrier who is not afraid to talk to you about methods and results.

Some things you should know about the care of the feet include: Draft horses that are shod for show purposes frequently have their feet trimmed and forced into artificial shapes which cause the animal to stand and move unnaturally, with an excessive false action. The result, after a few short years, is that these animals are highly prone to forms of navicular lameness, as excessive strain has been maintained on those tendons which anchor around the navicular bone near the heel of the foot. There are other forms of lameness which occur as a direct result of these unfortunate shoeing and trimming practices.

The ideal shape and angle of the hoof follows the natural angle of the pastern and the natural spread of

the quarters (see page 65). The foot should set flat on the ground and carry a straight line up through the leg. The foot should not toe in or out of the natural line of the leg. The shod foot should not prevent the legs from moving in a natural straight-ahead manner.

Some show people will argue until blue in the face against these precepts of natural hoof care. Unfortunately for many thousands of horses, or any of the animals with "forced" feet, the proponents of those practices cannot find sound horses, without artificial gait and lameness, to prove their approach as correct or even borderline acceptable.

These practices are not limited to the draft horse show world. One need only attend any gaited horse show to see the abominable extremes man employs to satisfy his fancy.

Some people prefer to use a "shoeing stock" which restricts the horse's movements. Care should be taken that such tools are not used to harm the animals and that the horses don't develop bad habits and refuse to allow their feet picked up without the stocks.

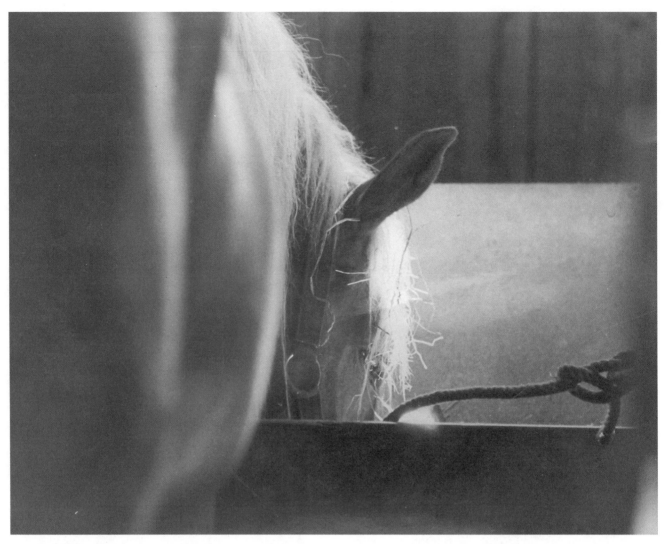

Photo by Nancy Roberts.

Cleaning The Feet. You will, at the very least, have to learn to pick up your horse's feet to be able to clean them occasionally. The feet will have to be cleaned regularly if the animal is walking in gravel and /or mud, so as to prevent thrush or stone problems.

With the front feet: stand facing the rear of the horse, but with your legs in a straight line with the horse's front legs. Do not stand ahead of or behind the horse's legs. Lean your shoulder into the horse's shoul- der and gently run your hand down the backside of the leg to the fetlock. Making sure that the animal is standing squarely on all four legs, or at least the other three, pick up the foot. Practice keeping your feet and legs out of the natural forward path of that leg you are holding up because, if the horse should pull it away from you, it will hurt. With a hoof-pick clean out the bottom, watching for bruises, infection and stones. The same procedure is used on the other front leg. The back legs are almost the same. Standing with your legs in line with the horse's legs, run your hand gently down the leg to the fetlock. Pick up the foot, always thinking about where you are in relationship to where the horse could go. Smoothly and quickly, when you have the leg up, step back and under the leg, bracing it with your own. With both the front legs and the back it is important that you do not spoil a horse because of what you do not know. Have your farrier around the first time you try this and ask for his or her help and guidance. That way you will be working towards positive lessons rather than failure.

Conclusion – There is a great deal of material in this chapter. But each of the many subjects has only been touched upon to give ideas and understanding, not to give the last word. With most of these subjects, the beginner will look on and elsewhere for additional material and that is as it should be.

CHAPTER FIVE

FEEDING THE WORK HORSE

*The late Willard Lee, of Oregon, works at cleaning up a dead furrow with his excellent
logging and farm team. Photo by Nancy Roberts*

Work horses must have proper feed, properly fed if they are expected to perform efficiently for a full lifetime. Unfortunately, less care is given the feeding of horses, used or not, than other classes of livestock. Considering the fact that savings of up to 20% in the cost of feeding can be realized by using care and intelligence in providing horses with feed, the person concerned with the viable application of horsepower should put feeding (maintenance) high on the list of priorities. Put another way, careless feed practices can result in greater expense, less work, unhealthy animals and bad feelings. You must be a good feeder if you are to be a good teamster, because it is an essential responsibility that affects performance.

There is a science to good feeding practice, and the comprehensive textbook of that science of feeding livestock has been Morrison's *Feeds and Feeding* since 1898. This book has been revised and re-issued in many editions up to recently. At the time of this writing it is out of print and information could not be secured as to whether or not recent editions contain specialized information on the feeding of work horses. The earlier editions did. Most of the information and many of the charts used in this chapter have been gleaned, paraphrased, and otherwise edited from *Feeds and Feeding, A Handbook for the Student and Stockman* by Frank B. Morrison, 21st edition, published by the Morrison Publishing Company, 1948. Additional information from other sources and from experience have also been used in this chapter.

required feed

Total Digestible Nutrients (TDN) or net energy, in liberal quantities, is the main requirement of a horse or mule doing muscular work. When working, the animal requires just a little more protein than while idle.

The same thing is true with regard to minerals and vitamins. So, surprisingly to many people, feed rations that are relatively low in protein and that have only a moderate supply of minerals and vitamins are entirely satisfactory for adult work horses.

Mature idle horses use two thirds of the consumed feed ration to supply body heat. The other third is all that is required to maintain bodily functions. So, without work, cheap feeds like straw or corn stalks, which are low in net energy and protein, are sufficient because they produce lots of body heat.

Pregnant brood mares require large amounts of protein, vitamins and minerals, and mares nursing foals require even greater amounts. The diet of young growing animals requires adequate amounts of protein, with concern for the intake of calcium, phosphorus and vitamins.

Horses differ greatly from other classes of farm stock, not only in temperament, but most importantly in feed requirements. And individual horses differ from one to another. For this reason, the charts and tables presented here serve only as a guideline. With experience you will find that one horse will need more feed while another requires less. But if an individual horse takes more and more feed and shows poor condition, look to the teeth, to internal parasites and to environment for causes.

A horse at hard work will take up to double the amount of TDN, but preferably close to the same amount of roughage as the idle horse. That means that more of the feed must consist of grains and/or other concentrates in order to achieve the desired balance. The reason that the amount of bulk fed should preferably remain constant is because of the horse's digestive system. The horse is not a ruminant, like a cow, and has a single stomach and a highly sensitive overall digestive system that is prone to irritations that fall under the heading of "colic." As described in Chapter Four, colic is a serious problem and as with any malady that might befall the horse, prevention is the very best insurance. And in the case of feeding, maintaining a consistent program without dramatic changes in quantity, character, or composition of feedstuffs is the

best policy. (The exception to this rule is, of course, the reduction of TDN when work is dramatically reduced or else serious conditions like Azoturia might result.)

The horse has a relatively small stomach and requires time to consume and digest adequate amounts of feed. Within this physical reality lies a clue to the best system of and approach towards feeding the work horse. (See METHODS OF FEEDING, this chapter.)

digestibility

Horses digest their feed less completely than do cattle or sheep, but the difference isn't great except when you look to low-grade roughages such as straw or poor quality hay. Moderate work, even immediately after eating, tends to increase digestion and absorption of nutrients. Severe labor may, however, slow digestion. Reasonable exercise does not waste nutrients by hastening their passage into the small intestine.

The digestion of grains and concentrates and their absorption into the system as nutrients can be increased if a degree of roughage is consumed by the horse before eating grain. The author has also found that slowing the process of eating increases digestion and absorption of nutrients (see METHODS OF FEEDING, this chapter).

proportions of roughages and concentrates

Hay or other roughage may be adequate for the maintenance of mature idle horses. On the other hand, horses at work need concentrates in addition to roughage and the proportions will vary depending on the severity of the work and the age and condition of the animal.

If hay is hard to come by, tests have proven that the amount of roughage can be reduced some. Horses at severe work have been maintained on as little as eight pounds of hay per head daily so long as they had a sufficient allowance of grain. In some cases, growing animals have been fed rations which included no hay but substituted such things as beet pulp for roughage (or fiber). These animals made satisfactory gains.

The opposite direction with regard to roughage proportions is a different story. Feeding too much hay or other roughage can be injurious to the animal. If given all the good hay they can eat, horses often eat too much, resulting in labored breathing and quick tiring. As mentioned before, horses have a relatively small

Three Belgians get a midday drink before returning to the field. Photo by Lynn Miller.

sized simple stomach that does not have capacity for great amounts of roughage all at once. Also, if the digestive tract is distended or swollen with too much roughage, the horse will be uncomfortable at hard labor and digestive disturbances could result.

Therefore, instead of the popular practice of keeping mangers full, the allowance of hay should be limited. Not only is this a good way to save feed, but it is also beneficial to the horses. Limiting the amount of hay available at any one time is especially important when feeding rich, palatable hay like alfalfa, clover or early cut timothy.

A probable conclusion from mixed data had 2.5 to 3 lbs. of hay (or more) equaling 1 lb. of corn or oats for the production of work. For maintaining an idle horse in winter, hay will have a higher value and it will probably take only 1.5 to 2 lbs. of hay to equal 1 lb. of grain in value.

protein requirements

Morrison's feeding standards recommends 0.6 to 0.8 lbs. of digestible protein per day for a 1,000 lb. idle horse. Of course, when the horse is at work, somewhat larger amounts are recommended. There are, however, other standards which frequently call for greater amounts of digestible protein and some tests the result of which indicate that less protein does not harm the animal. There seems to be some correlation between the amount of protein in a ration and its digestibility suggesting that a low protein content reduces digestion. Also, it should be pointed out that young growing horses and pregnant and nursing mares require more protein than do other classes of horses.

mineral requirements

The calcium and phosphorus needs of mature work horses are not great. These minerals are essential in adequate supply for brood mares and young developing animals and to a lesser degree even with mature work horses. It is possible to get all the calcium and phosphorus that is required from good to excellent quality hay. But, even though the mineral needs are not high, an unusually low level of calcium and phosphorus in the diet can result in serious bone damage, even for mature horses. It is possible, for a low fee, to have agricultural stations or private agricultural laboratories do a tissue analysis of hay to determine conclusively the exact mineral and protein content. The expense is well warranted as it may save unnecessary expenditures on either vet bills or for unneeded mineral supplements.

Growing foals require an ample supply of not only calcium and phosphorus but also vitamin D to enable them to slowly develop strong, sound bones. It is this author's contention that if the development and growth of the bones are forced too fast it results in substantially weaker and inferior bone.

Some of the effects that may come from a deficiency of calcium and phosphorus include; being run down, emaciated, lame, reducing quality of stallion semen, bone disease.

In areas where there is a deficiency of iodine, brood mares should be fed iodized salt during at least the latter half of their pregnancy to avoid health problems with the newborn foals.

The soils in different parts of this country may have marked deficiencies of certain minerals. Local agricultural authorities will have information about any mineral deficiencies in the soil and that information should be used to guide the individual in amending rations to include necessary trace minerals. On the good mixed farm where the best cropping and fertilizing practices are employed, the roughages and grain will be naturally mineral and vitamin rich and less deficiency-related problems will be seen. The teamster who must purchase all or any significant portion of the horse's ration off the farm should exercise care in acquiring all the needs of the animal. A good plan would be to locate the better farmers in the area and purchase off the farm the roughages and grains needed rather than to rely on mass-produced premix rations.

salt

Horses must have a regular supply of salt. An allowance of 1.75 to 2.0 ounces per head daily is sufficient and many horses eat less. A good program is to provide free choice salt either loose in a box or in block form. In recent years, block and loose salt has been pre-mixed with trace minerals, iodized and even mixed with a fly repellent which works to repel flies after it is sweated out onto the horse's hide with salt. Horses at hard work need more salt than others as they lose a lot when they sweat.

vitamins

Morrison's Feeds and Feeding reports that there are commonly no deficiencies of vitamins for mature horses in ordinary rations that contain plenty of good hay, for their vitamin requirements are small. The vitamin needs of brood mares and foals are higher, but they will be amply met if good legume hay or mixed hay is fed when the horses are not on pasture.

The requirements of work horses for vitamin A (or carotene) per unit of body weight are about the same as those for cattle, sheep or hogs. A deficiency of vitamin A could cause night blindness, eye injury, respiratory and reproductive difficulties and eventually death. Foals sometimes develop rickets from a lack of vitamin D or of calcium and phosphorus.

The usual rations fed horses, containing plenty of good hay will ordinarily provide all of the B-complex vitamins they require. But because of the nature of the horse's digestive system, it is important that they be provided adequate amounts. Without this complex, normal growth cannot be expected.

Moonblindness, or periodic opthalmia, can be prevented, but not cured, by adding 40 milligrams of riboflavin per head to the daily ration.

vitamin and mineral supplements

There are many commercially prepared vitamin/mineral/enzyme/micronutrient/trace mineral type premixes on the market. Some are even advertised to be of specific value to horses and three or four actually claim to be formulated with the work horse in mind. The advertising claims of effectiveness are aimed at two sensitive areas of justified concern throughout the entire horse industry. Those concerns are for the rehabilitation of horses in poor condition (there are many – mainly the backyard inmate) and for improving the increasingly poor fertility of not only mares but stallions as well. Vitamins and minerals in adequate supply are important to the health of the work horse but they will not correct problems such as internal parasite infestation or inadequate feed. As for the reproductive system, stallions should not be kept as such and used for breeding unless they are superior representatives of their breed and/or type and unless they are found to be free of genetic defects or characteristics which will be detrimental to the future. Unfortunately, horses will continue to be abused and misunderstood by man. Economics seems the one area where we should make inroads towards changing management habits; but people continue to purchase expensive "cures" and "wonder tonics & potions" for problems that can be prevented with adequate feed, common sense and much less money.

This author has used very expensive vitamin/mineral premixes on Belgian work horses including a stallion in service, open and bred mares, young stock and working geldings, for a period of one year. For six years, this author has used no mineral mix, just home-grown hay and grain plus iodized salt and occasionally molasses or apples as treats mixed with the oats. The result for this author was NO DIFFERENCE. There was absolutely no recognizable difference in physical condition or performance.

Certainly there may be specific instances where the expensive premix mineral rations are of constructive value to a particular individual animal.

All the necessary nutrients, minerals, vitamins and even micronutrients can be provided in a well-balanced ration of good quality hay and grain. And if there is a specific soil deficiency (of, say, selenium, boron or some other trace mineral), that can be remedied by purchasing just the missing or deficient mineral (or vitamin) and including it in the feed ration. In this manner, you will be saving money over the cost of premixed minerals. In 1976, the author spent $300 in one year for the mineral supplement referred to above, for 10 horses.

The intelligent procedure to follow would be to first inquire of your local government soils office to find if there is any mineral deficiency in your specific soil type. Then have your soil, or the soil where your feed will be grown, tested. And finally have the forage and even the grain tested at your local agricultural college laboratory to determine what it contains and lacks. From the tables contained in this chapter and from other reference sources you should be able to compute your animal's requirements for specific vitamins and minerals. You should be surprised to find that most, if not ,all, will be supplied amply by the natural ration. If there is a discovered deficiency check with the feed store, the veterinarian and the ag college about sources for the ingredient and compare the price and information. Now compare the cost of providing this missing ingredient with the cost of feeding a "special" premix. A mix will probably provide quite a few unnecessary, unneeded and/or expensive vitamins and minerals.

watering horses

Horses must have plenty of good quality water. It could take 10 to 12 gallons or more of water per horse per day. In warm weather and when at hard work, horses will drink more water than at other times: that is simply because of the evaporation of water from their bodies. Curiously, horses will drink more water when on a diet of rich legume hay rather than straight grass hay.

When horses drink water, it doesn't appear to have any bearing on digestion. It is, however, important to keep the animals on a regular schedule of access to water if it is not free choice. A horse that has been worked hard should be watered before being fed, especially grain, but HE SHOULD NOT BE AL-LOWED MUCH WATER WHEN VERY WARM. The best practice is to give the animal a short drink and allow him to cool a little while eating roughage and then to allow another drink before the grain ration. For decades good horsemen chose to err on the side of caution when it came to allowing working horses free access to water. There are new findings and new semi-official conclusions on this subject.
The old rule went something like this:

A HOT HORSE GIVEN FREE CHOICE COOL WATER COULD DEVELOP A COLIC CONDITION. So allow a limited drink when first coming in from the field.

A HOT HORSE GIVEN FULL GRAIN AND HAY RATION FIRST AND THEN AL-LOWED FREE CHOICE TO WATER WILL GET COLIC!

During exceedingly hot weather horses should be watered every hour or two while at hard work. They feel like you do when working in the heat and a short drink will help to refresh them. Taking a can or barrel of water to the field, if farming, might save you, in 100 degree heat, from having animals overheat and get sick.
The new rule:

THERE IS NO EVIDENCE TO CON-CLUDE THAT FREE CHOICE WATER, AT ANY TIME, WILL CAUSE COLIC IN HORSES. ALLOW THEM ALL THEY WILL DRINK.

CAUTION SHOULD STILL BE EXER-CISED WHEN ALLOWING HORSES TO EAT AND THEN DRINK.

In spite of this information, this author, after 30 plus years of working horses, can't seem to break his cautionary habits. So I continue to follow the old rule.

For young and old alike, pasture can be a real tonic. Photo by Kristi Gilman-Miller.

FEEDS

Commonly, across the U.S., horse rations usually consist of one or two grains and one or maybe two roughages. Oats are by far the most common, and most well thought of, grain for horses. They are certainly the safest grain as they have a good balance of nutrients and the bulk and fiber necessary for correct digestion. But even though many horsemen consider oats to be indispensable, satisfactory results are had with grain mixtures that contain no oats, but are properly balanced and have adequate bulkiness.

A primary consideration in making up or deciding on the grain or concentrate ration for the work horse should be cost plus availability.

It is critically important for horses that their feeds not be moldy, spoiled or extremely dusty. Damaged feeds will cause problems with the digestive tract and system and dusty feeds will cause serious problems with the respiratory works.

HAY – When horses were the power mainstay in this country, timothy hay was the standard by which all other roughages were measured. The major reasons for the popularity of timothy hay were that it is a safe feed, being free of dust and less likely to spoil and that it provided, if of good quality, adequate protein, mineral and vitamins. As horses fell from popularity so did timothy and farmers turned to grasses and grass-legume mixtures which were suitable for a variety of livestock. Well-cured legume-grass or straight legume hay is entirely satisfactory if fed properly. You do have to be careful that the hay is properly cured and that

horses are not given free choice because they will eat themselves sick. Legume hay is higher in total digestible nutrients and so less should be fed than straight grass hay.

SILAGE – This author is wary of feeding silage to work horses due to the sensitivity of the digestive system and the nature of silage. Morrison's *Feeds and Feeding* say that poor quality silage should never be fed horses but that good quality silage, free of mold or decay, can be 1/3 to 1/2 of the roughage ration. They also say that horses should be gradually accustomed to silage and that horses at hard work should not be fed much silage.

GRAINS – Oats are unexcelled in feeding horses but corn, wheat, barley, and the grain sorghums can all be used successfully in place of oats when fed carefully as either an appropriate mixture or grind. Corn is best fed in the form of ear corn or shelled corn. The other grains mentioned should be crushed or ground and should be mixed with some bulky feed to avoid danger of colic. Ground or crushed rye should not be more than one-third of the concentrate mixture. Hominy feed is a satisfactory substitute for oats and molasses is sometimes fed to horse as an appetizer or conditioner.

Wheat bran is an excellent supplement because of its bulky nature and laxative effect on horses. Linseed meal also has a laxative effect with the added feature of being a good conditioner.

PASTURE – Availability of good pasture for work horses can save a great deal of feed costs and has a rejuvenating effect on the animals as well. Pasture is especially important for foals and brood mares.

When horses are worked regularly, if pasture is available, the animals should be turned out after they have eaten a little roughage and all of their evening grain ration. The pasture will help keep the animals fresh and spry.

Be careful with early spring grass as it can cause problems with diarrhea and even colic. It is best to get the animal slowly on to the spring grass, feeding plenty of hay and allowing short but regular visits to pasture for a few days increasing the amount of time spent on the lush grass until the animal can be left out. Even then it is a good idea to provide some dry hay to slow the laxative effect of the new grass. If horses are being worked and have access to spring pasture NO other laxative type feed should be used, as the effect of

A young pregnant Belgian mare on pasture.

the grass will be more than ample in this respect.

In the fall, after the farm work is completed, if the teamster has access to crop land with residues such as corn stalks, the horses can be turned out to clean up the waste as long as some hay is available. Care should be taken to make sure that growing foals and brood mares are receiving adequate nutrients in any season. Horses used extensively on the highway should be turned out on pastures for spells to allow their feet to recover from the concussive effect of the pavement.

Mixed pastures which contain a considerable measure of white clovers or other legumes are the very best for horses. The use of pastures grown in a regular crop rotation, instead of parasite-infested permanent pasture, will greatly help in reducing worm problems in the work horse. Since horses are less susceptible to bloat than cattle or sheep, alfalfa or other legume pastures are excellent for them. The best and safest results will always be had from pastures which combine grasses and legumes.

methods of feeding

This author's experience and observations have led to the conviction that good quality feed in adequate portions is only half the battle in maintaining work horses in good condition. How the animal is fed can have the determining factor in efficiency. In this regard it is up to the care and intelligence of the individual teamster whether success will be a regular visitor or a casual passerby. But it should be said that good feeding practices need not be complex or difficult, in truth better feeding will ultimately be easier and save time and money.

The teamster, of course, will have to make the decisions of how much to feed to which class of animal. As stated, the amount of feed necessary for a work horse will depend on the size of the animal, on the severity of the work and also on whether he is an easy keeper or a hard one. As a rule of thumb to guide you: allow a total of 2 to 2.5 lbs. daily of concentrates and roughage combined per 100 lbs. of live weight. The charts and tables contained in this chapter break that down further. You will have to learn to recognize the "easy keepers" from the "hard keepers" and adjust individual rations so that the fat horse is not getting fatter and the other is up in full shape. The only way to learn this is by looking, paying attention and using common sense.

Photo by Kristi Gilman-Miller.

Usually the daily amount of grain or other concentrates is divided equally into three feeds and given at morning, noon and night. It is best to feed only a small amount of hay in the morning, so the horse's digestive tract will not become distended too much when he is at work. With hay, the common plan is to feed one-fourth the ration in the morning, one-fourth at noon and the balance (1/2) in the evening. This author prefers to do the same sort of split with the grain ration as well because the horses have more time to eat in the evening and because the absorption of nutrients will be more complete than if the animal's metabolism and digestive process is being affected by strenuous labor.

The slower a horse eats, or the longer it takes for the animal to digest its feed, the more value will be derived. If you allow your horse to bolt his grain you are doing him and yourself a disservice. You can make the animal eat the grain slower by placing round flat stones three inches or wider in the grain box. Small salt blocks can also be used for this purpose. The same

thing might be accomplished with corn cobs. A popular plan in England is to mix the grain with chaff or chopped roughage.

To follow suit, it is a good practice not to allow horses access to both hay and grain the minute they arrive at the manger. Their natural inclination will be to consume the grain quickly first and then take their time eating the hay. The best practice is to have them come to the manger with only hay in front of them. Allow them to eat a little hay, then place the grain in their feed boxes. The roughage in the digestive tract will slow the digestion of the grain and make for better utilization.

To avoid digestive problems and possible deaths from *azoturia* (see Chapter 4), the allowance of grain for horses at hard work MUST BE REDUCED on idle days to 50 to 70 percent of the amount usually fed. It is best to feed on such days, in place of the grain, a mixture of two-thirds grain and one-third bran. Some feed a small allowance of grain at noon on idle days, with only a bran mash both morning and night.

If the animals are working full days, an hour should be allowed for the noon meal. This author prefers to allow two hours and work a little later than usual as it helps the horses to finish the days work in better condition. If working during extreme heat it may prove valuable to work very early and take several hours off midday, finishing the days work in the cool of the evening.

When the horses come in at the end of the work day they should be given a short drink of water, be unharnessed, allowed access to hay, allowed another drink of water, given hay and grain ration, and when sweat has dried, be brushed well. If the animals are to be stabled all night they should, well after finishing the main ration, be allowed to drink all the water they wish.

WINTERING FARM HORSES – If horses are to be idle in the winter they can and should be maintained on the cheaper roughages available. If the horses are expected to get by on just corn stover or straw, five pounds of good quality hay should be provided to each horse three times a week. Light grain feeding should begin before the heavy work season is anticipated, such as the spring. Idle horses should be able to exercise freely and in the winter, if running on pasture, should be provided some sort of shelter break to protect them from the worst weather. Be sure to provide free choice salt and water.

feeding the breeding unit

BROODMARES – It can be profitable to keep a good team of broodmares to do part or all of the work required and also raise a pair of foals each year. A brood mare does not require very much more feed than a gelding doing equal work.

Mares are less fertile than other farm animals and the conception rate is rarely over 60 to 70% (often lower). Unless the mare is properly fed and cared for a good reproductive success cannot be expected.

Broodmares must have plenty of exercise to keep their muscle tone and attitude right. For this reason working the broodmare is an excellent practice and it contributes to their general health. But care must be taken not to cause the mare to pull too hard, especially where the going is slippery. Stay away from deep mush or snow and be careful in backing heavy loads. Take every precaution to prevent the mare from slipping and/or falling, especially when pulling. Mares can work in harness up to three days before foaling without hardship. How soon they are put back to work will depend on their condition after foaling.

The feed rations of brood mares should contain liberal amounts of protein, calcium, phosphorus and vitamins. Young growing mares and those suckling foals will require the most as the charts illustrate. Good quality grass-legume hay should provide adequate amounts of the required nutrients. Less than ideal roughage can be offset by feeding grain or concentrates. Care should be taken not to allow the broodmare, settled or open, to get too fat.

The average gestation period for work mares is 11 months or 335 days, but that can vary considerably. Shortly before foaling, the grain allowance should be decreased and enough bran or other laxative feeds used to prevent constipation. Unless the mare can safely and comfortably foal in an empty small pasture, the best foaling setup is a roomy, disinfected, well-bedded box stall (10' x 10' is too small, 14' x 14' is better). Before foaling, the horse owner should know where a competent veterinarian can be reached and what signs to look for as indications of difficulty for the mare.

The mare should be given a half bucket of lukewarm water before foaling, if she will readily drink it. After foaling and back on her feet she will need another drink. A light bran feeding after foaling is best and this may be followed by oats or oats and bran.

Beginning two or three days after foaling, the mare and foal should be given some exercise each day, but be careful not to put them with other horses, as fighting and injury will probably result. If the mare has no trouble foaling she can do light work in harness within a week. If she has had trouble, allow a longer rest.

Farm mares should not usually be bred until they are three years old. Breeding a well-grown and cared for two-year-old filly will not set her back but care should be taken to provide all of her nutritional needs.

THE FOAL – To reach good size at maturity, a foal should make about half its entire growth during the first year. This means that it will have to be fed liberally. Soon after birth the foal should receive the mare's colostrum (or first milk), because this increases the resistance to disease and infection and furnishes critical vitamin A. The foal's navel should be carefully disinfected with tincture of iodine.

Care should be taken to provide the kinds and quantity of feeds that will stimulate and maintain good milk production in the mare. Good pasture is best. If it is not available, good quality legume-grass hay with an allowance of grains will suffice. It is possible for the milk flow of the mare to be too great and the result may be indigestion of the foal. Cutting back on the feed of the mare will regulate this infrequent problem.

THE STALLION – The most important thing to the well-being of a stallion is ample exercise and the best exercise is work. Also the best advertisement of a stallion is the sight of him at work. It is a bad plan to restrict a stallion to a box stall. He should at least have free access to a paddock area.

The stallion should be fed good quality roughages and grain, and his ration, during the breeding season, should equal that of a horse at hard work.

FEEDING FOALS – Foals should learn to eat grain as early as possible. By the time they are weaned, foals should be eating two to three pounds of concentrates per day. This should go along with good quality hay and free choice water.

If the mares and foals are on pasture, a small fenced-off area, set up as a "creep feeder" can be provided with a gate just big enough for the foals to enter and help themselves to grain, but small enough to keep the mares out. One way to get the foals used to this arrangement is to let the mares in for the first few days with the youngsters. After the mares are locked out make sure to provide a salt lick to keep them nearby while the foals are eating.

At between five and six months, the foal should be weaned from the mare. To wean, the foal should be removed from the mare and not allowed back with her until they have both forgotten each other. The grain allowance of the mare should be slowly slacked off.

After weaning, the foals should be kept growing thriftily with liberal feed rations carefully calculated to provide calcium, phosphorus and vitamin needs. Unfortunately, the current boom in draft horse prices and the fetish for "big" horses at any cost has many farmers feeding heavy protein-rich rations, forcing foals to excessive gains without concern for adequate exercise and mineral/vitamin needs. The result is a great many big beautiful two and three year old draft horses with poor bone and muscle tone. These animals do poorly when put to the test of long days at hard work. Some of the better modern day horse feeders feel that it is more important to grow bone, height and proper muscle structure before any concern for "fat."

FEED AMOUNTS

Work horses and mules should be fed approximately the following amounts of grain and hay per 100 lbs. of animal live weight:

At hard work – 1 to 1.4 lbs. of grain, 1 lb. of hay
At medium work – .75 to 1 lb. of grain, 1 to 1.25 lbs. of hay
At light work – .4 to .75 lb. of grain and 1.25 to 1.5 lbs. of hay
Idle – Chiefly on roughage with grain if roughage is poor quality

The standards for horses at hard work are actually working for 7 to 8 hours per day.

VITAMIN CONTENT OF FEEDING STUFFS

Feeding stuff	Caro-tene	Vita-min A-activity	Thia-min	Ribo-flavin	Nia-cin	Panto-thenic acid
	mg. per lb.	I. U. per lb.	mg. per lb.	mg. per lb.	mg. per lb.	mg. per lb.
Alfalfa hay, all analyses	11.4	19,000	1.3	6.2	17.4	8.1
Clover-and-grass mixed hay, mostly clover, U. S. grade no. 3	3.2	5,333				
Corn, dent, yellow	2.2	3,667	1.9	0.5	9.0	2.3
Oats, grain	0.05	83	2.8	0.5	6.3	6.0
Wheat bran	0.08	133	3.9	1.4	63.5	13.6

AVERAGE COMPOSITION AND DIGESTIBLE NUTRIENTS
in percentages

Feeding stuff	Total dry matter	Dig. protein	Total dig. nutri-ents	Average total composition			
				Protein	Fat	Fiber	Mineral matter
Dry Roughages							
Alfalfa hay, good (28-31% fiber)	90.5	10.3	50.4	14.3	1.8	29.7	8.2
Alfalfa and timothy hay	89.8	6.6	49.1	11.1	2.2	29.5	6.7
Bluegrass hay, native western	91.9	6.7	52.6	11.2	3.0	29.8	8.0
Clover, Lodino, and grass hay	88.0	11.1	53.3	16.3	2.2	20.7	7.1
Clover and mixed grass hay, high in clover	89.7	5.5	52.2	9.6	2.7	28.8	6.2
Corn stalks, dried	82.8	0.8	40.7	4.7	1.5	28.0	5.3
Mixed hay, good, more than 30% legumes	88.0	5.2	50.9	9.2	1.9	28.1	6.0
Oat hay	88.1	4.9	47.3	8.2	2.7	28.1	6.9
Prairie hay, western, good	90.7	2.1	49.6	5.7	2.3	30.4	7.4
Timothy hay, before bloom	89.0	5.4	56.8	9.7	2.7	27.4	6.5
Vetch and oat hay, over ½ vetch	87.6	8.4	52.7	11.9	2.7	27.3	8.2
Silages							
Corn, dent, well-matured, well-eared	28.4	1.3	20.0	2.3	0.9	6.3	1.6
Concentrates							
Barley feed, high grade	90.3	10.8	73.2	13.5	3.5	8.7	4.1
Beet pulp, dried	90.1	4.3	67.8	9.2	0.5	19.8	3.4
Buckwheat, ordinary varieties	88.0	7.4	62.2	10.3	2.3	10.7	1.9
Corn, dent, grade no. 1	87.0	6.8	82.0	8.8	4.0	2.1	1.2
Corn ears, including kernels and cobs (corn and cob meal)	86.1	5.3	73.2	7.3	3.2	8.0	1.3
Linseed meal, o.p. 37% protein or more	90.9	33.1	77.4	38.0	5.9	7.7	5.6
Molasses, cane, or blackstrap	74.0	0	54.0	2.9	0	0	9.0
Oats, Pacific coast states	91.2	7.0	72.2	9.0	5.4	11.0	3.7
Wheat, average of all types	89.5	11.1	80.0	13.2	1.9	2.6	1.9
Wheat bran, all analyses	90.1	13.7	67.2	16.9	4.6	9.6	6.1

North Dakota's Odegaard Belgians photographed by Fuller Sheldon.

AVERAGE COMPOSITION AND DIGESTIBLE NUTRIENTS
in percentages

Feeding stuff	Mineral and fertilizing constituents				Digestion coefficients		
	Calcium	Phos-phorus	Nitro-gen	Potas-sium	Protein	Fat	Fiber
Dry Roughages							
Alfalfa hay, good (28-31% fiber)	1.27	0.22	2.29	2.01	72	34	43
Alfalfa and timothy hay	0.81	0.21	1.78	1.78			
Bluegrass hay, native western			1.79				
Clover, Lodino, and grass hay	1.05	0.26	2.61	1.97			
Clover and mixed grass hay, high in clover	0.90	0.19	1.54	1.46	57	47	61
Corn stalks, dried	0.25	0.09	0.75	0.50			
Mixed hay, good, more than 30% legumes	0.90	0.19	1.47	1.46			
Oat hay	0.21	0.19	1.31	0.83	60	65	51
Prairie hay, western, good	0.36	0.18	0.91		37	38	64
Timothy hay, before bloom			1.55		56	36	75
Vetch and oat hay, over ½ vetch	0.76	0.27	1.90	1.51	71	52	51
Silages							
Corn, dent, well-matured, well-eared	0.08	0.06	0.37	0.27	55	80	65
Concentrates							
Barley feed, high grade	0.03	0.40	2.16	0.60			
Beet pulp, dried	0.67	0.08	1.47	0.18	47	0	75
Buckwheat, ordinary varieties	0.09	0.31	1.64	0.45	72	80	45
Corn, dent, grade no. 1	0.02	0.28	1.41	0.28	77	90	57
Corn ears, including kernels and cobs (corn and cob meal)		0.22	1.17	0.29			
Linseed meal, o.p. 37% protein or more	0.39	0.86	6.08	1.10			
Molasses, cane, or blackstrap	0.74	0.08	0.46	3.67	0		
Oats, Pacific coast states			1.44				
Wheat, average of all types	0.04	0.39	2.11	0.42	84	81	70
Wheat bran, all analyses	0.14	1.29	2.70	1.23	81	83	49

A PTO haybine is propelled by a motorized forecart and a four abreast. PA Horse Progress Days 2000. Photo by Lynn Miller.

MORRISON FEEDING STANDARDS

	Requirements per head daily								
	Dry matter	Diges-tible protein	Total digestible nutrients	Calcium		Phosphorus		Caro-tene	Net energy
weight	lbs.	lbs.	lbs.	grams	lb.	grams	lb.	mg.	therms
Horses or mules, idle									
1,000 lbs.	13.0-18.0	.6- .8	7.0- 9.0	15.0	.033	15.0	.033	55	(5.6- 7.2)
1,100 lbs.	13.9-19.3	.7- .9	7.5- 9.7	16.5	.036	16.5	.036	61	(6.0- 7.7)
1,200 lbs	14.8-20.6	.7- .9	8.0-10.3	18.0	.040	18.0	.040	66	(6.4- 8.2)
1,300 lbs.	15.7-21.8	.7-1.0	8.5-10.9	19.5	.043	19.5	0.43	72	(6.8- 8.7)
1,400 lbs.	16.6-23.0	.8-1.0	8.9-11.5	21.0	.046	21.0	.046	77	(7.2- 9.2)
1,500 lbs.	17.5-24.2	.8-1.1	9.4-12.1	22.5	.050	22.5	.050	83	(7.5- 9.7)
1,600 lbs.	18.3-25.4	.8-1.1	9.9-12.7	24.0	.053	24.0	.053	88	(7.9-10.1)
1,700 lbs.	19.1-26.5	.9-1.2	10.3-13.3	25.5	.056	25.5	.056	94	(8.2-10.6)
1,800 lbs.	20.0-27.6	.9-1.2	10.8-13.8	27.0	.060	27.0	.060	99	(8.6-11.1)
Horses or mules at light work									
1,000 lbs.	15.0-20.0	.8-1.0	9.0-11.0	15.0	.033	15.0	.033	55	7.5- 9.1
1,100 lbs.	16.2-21.6	.9-1.1	9.7-11.9	16.5	.036	16.5	.036	61	8.1- 9.8
1,200 lbs.	17.4-23.1	.9-1.2	10.4-12.7	18.0	.040	18.0	.040	66	8.7-10.5
1,300 lbs.	18.5-24.7	1.0-1.2	11.1-13.6	19.5	.043	19.5	.043	72	9.3-11.2
1,400 lbs.	19.6-26.3	1.0-1.3	11.8-14.4	21.0	.046	21.0	.046	77	9.8-11.9
1,500 lbs.	20.8-27.7	1.1-1.4	12.5-15.2	22.5	.050	22.5	.050	83	10.4-12.6
1,600 lbs.	21.9-29.2	1.2-1.5	13.1-16.0	24.0	.053	24.0	.053	88	10.9-13.3
1,700 lbs.	23.0-30.6	1.2-1.5	13.8-16.8	25.5	.056	25.5	.056	94	11.5-13.9
1,800 lbs.	24.0-32.0	1.3-1.6	14.4-17.6	27.0	.060	27.0	.060	99	12.0-14.6
Horses or mules at medium work									
1,000 lbs.	16.0-21.0	1.0-1.2	11.0-13.0	15.0	.033	15.0	.033	55	9.4-11.1
1,100 lbs.	17.4-22.8	1.1-1.3	11.9-14.1	16.5	.036	16.5	.036	61	10.2-12.1
1,200 lbs.	18.8-24.6	1.2-1.4	12.9-15.2	18.0	.040	18.0	.040	66	11.0-13.0
1,300 lbs.	20.1-26.4	1.3-1.5	13.8-16.3	19.5	.043	19.5	.043	72	11.8-14.0
1,400 lbs.	21.5-28.2	1.3-1.6	14.8-17.4	21.0	.046	21.0	.046	77	12.6-14.9
1,500 lbs.	22.8-29.9	1.4-1.7	15.7-18.5	22.5	.050	22.5	.050	83	13.4-15.8
1,600 lbs.	24.1-31.6	1.5-1.8	16.6-19.6	24.0	.053	24.0	.053	88	14.2-16.7
1,700 lbs.	25.4-33.3	1.6-1.9	17.5-20.6	25.5	.056	25.5	.056	94	14.9-17.6
1,800 lbs.	26.7-35.0	1.7-2.0	18.3-21.7	27.0	.060	27.0	.060	99	15.7-18.5
Horses or mules at hard work									
1,000 lbs.	18.0-22.0	1.2-1.4	13.0-16.0	15.0	.033	15.0	.033	55	11.3-13.9
1,100 lbs.	19.7-24.0	1.3-1.5	14.2-17.5	16.5	.036	16.5	.036	61	12.4-15.2
1,200 lbs.	21.3-26.1	1.4-1.7	15.4-19.0	18.0	.040	18.0	.040	66	13.4-16.5
1,300 lbs.	23.0-28.1	1.5-1.8	16.6-20.5	19.5	.043	19.5	.043	72	14.5-17.8
1,400 lbs.	24.7-30.2	1.6-1.9	17.8-21.9	21.0	.046	21.0	.046	77	15.5-19.1
1,500 lbs.	26.3-32.2	1.8-2.0	19.0-23.4	22.5	.050	22.5	.050	83	16.5-20.3
1,600 lbs.	28.0-34.2	1.9-2.2	20.2-24.8	24.0	.053	24.0	.052	88	17.5-21.6
1,700 lbs.	29.6-36.2	2.0-2.3	21.4-26.3	25.5	.056	25.5	.056	94	18.6-22.9
1,800 lbs.	31.2-38.1	2.1-2.4	22.5-27.7	27.0	.060	27.0	.060	99	19.6-24.1

Brood mares nursing foals, but not hard at work										
1,000 lbs.	15.0-22.0	1.2-1.5	9.0-12.0	35.0	.077	29.0	.064	70	7.6-10.0
1,100 lbs.	16.2-23.8	1.3-1.6	9.7-13.0	36.0	.079	30.0	.066	77	8.2-10.8
1,200 lbs.	17.4-25.5	1.4-1.7	10.4-13.9	37.0	.081	31.0	.068	84	8.8-11.6
1,300 lbs.	18.5-27.1	1.5-1.9	11.1-14.8	38.0	.084	32.0	.071	91	9.4-12.3
1,400 lbs.	19.6-28.8	1.6-2.0	11.8-15.7	39.0	.086	33.0	.073	98	10.0-13.1
1,500 lbs.	20.8-30.4	1.7-2.1	12.5-16.6	40.0	.088	34.0	.075	105	10.5-13.8
1,600 lbs.	21.9-32.1	1.7-2.2	13.1-17.5	41.0	.090	35.0	.077	112	11.1-14.6
1,700 lbs.	23.0-33.7	1.8-2.3	13.8-18.4	42.0	.093	36.0	.079	119	11.6-15.3
1,800 lbs.	24.0-35.2	1.9-2.4	14.4-19.2	43.0	.095	37.0	.081	126	12.2-16.0
Growing colts, after weaning									
400 lbs.	9.2-11.3	.8-.9	5.6-7.2	40.0	.088	30.0	.066	24	4.9-6.3
500 lbs.	10.9-13.3	.9-1.0	6.6-8.4	40.0	.088	30.0	.066	30	5.7-7.3
600 lbs.	12.4-15.2	1.0-1.2	7.6-9.6	40.0	.088	30.0	.066	36	6.5-8.3
700 lbs.	13.9-17.0	1.1-1.3	8.5-10.8	40.0	.088	30.0	.066	42	7.3-9.3
800 lbs.	15.3-18.7	1.2-1.4	9.4-11.9	40.0	.088	30.0	.066	48	8.0-10.1
900 lbs.	16.7-20.4	1.3-1.5	10.2-13.0	40.0	.088	30.0	.066	54	8.7-11.0
1,000 lbs.	18.0-22.0	1.4-1.6	11.0-14.0	35.0	.077	26.0	.057	60	9.2-11.8
1,100 lbs.	19.3-23.6	1.5-1.6	11.8-15.0	30.0	.066	23.0	.051	66	9.9-12.6
1,200 lbs.	20.6-25.1	1.5-1.7	12.6-16.0	30.0	.066	23.0	.051	72	10.6-13.4

MINERAL MATTER CONTENT

Feeding stuff	Calcium	Phosphorus	Potassium	Sodium	Chlorine	Sulfur	Magnesuim	Iron	Manganese	Copper
	%	%	%	%	%	%	%	%	mg. per lb.	mg. per lb
Alfalfa hay, all analyses	1.47	0.24	2.05	0.13	0.37	0.32	0.29	0.025	20.5	3.7
Clover and mixed grass hay, high in clover	0.90	0.19	1.46	0.17	0.64	0.13	0.25	0.022	42.1	3.2
Oat hay	0.21	0.19	0.83	0.15	0.46		0.16	0.049	36.6	
Prairie hay, western, good	0.36	0.18					0.25			
Timothy and clover hay, ¼ clover	0.51	0.20	1.48	0.17	0.59	0.13	0.17	0.017	36.6	2.5
Barley, common	0.06	0.37	0.49	0.06	0.15	0.15	0.13	0.008	8.0	5.8
Corn, dent, grade no. 1	0.02	0.28	0.28	0.01	0.06	0.12	0.10	0.003	2.6	1.8
Linseed meal, 37% protein or more	0.39	0.86	1.10	0.06	0.04	0.42	0.60			
Oats, grain	0.09	0.34	0.43	0.09	0.12	0.21	0.14	0.007	19.9	3.8
Wheat grain, average of all types	0.04	0.39	0.42	0.06	0.08	0.20	0.14	0.006	19.9	3.7

HORSES AND MULES

Horses and mules at hard work, weight 1,200 lbs.

1. Grass hay, 12 lbs.; oats, 16 lbs.
2. Grass hay, 12 lbs.; corn, 13 lbs.; linseed meal or other high-protein supplement, 1 lb.
3. Legume hay, 12 lbs.; corn, 13.5 lbs.
4. Legume hay, 6 lbs.; grass hay, 6 lbs.; corn, 14 lbs.
5. Shredded corn fodder, 6 lbs.; legume hay, 6 lbs.; oats, 15 lbs.
6. Oats or barley straw, chopped, 4 lbs.; legume hay, 8 lbs.; oats, 16 lbs.

Horses and mules at medium work, weight 1,200 lbs.

1. Grass hay, 14 lbs.; oats, 11 lbs.
2. Grass hay, 14 lbs.; corn, 9 lbs.; linseed meal or other high-proteien supplement, 0.75 lbs.
3. Legume hay, 14 lbs.; corn, 9 lbs.
4. Legume hay, 7 lbs.; grass hay, 7 lbs.; corn, 9.5 lbs.
5. Shredded corn fodder, 7 lbs.; legume hay, 7 lbs.; oats, 10 lbs.
6. Oat or barley straw, chopped, 5 lbs.; legume hay, 9 lbs.; oats, 11 lbs.

Horses and mules at light work, weight 1,200 lbs.

1. Grass hay, 16 lbs.; oats, 6 lbs.
2. Grass hay, 16 lbs.; corn, 4.5 lbs.; linseed meal or other high-protein supplement, 0.5 lbs.
3. Legume hay, 16 lbs.; corn, 4 lbs.
4. Legume hay, 8 lbs.; grass hay, 8 lbs.; corn, 4.5 lbs.
5. Shredded corn fodder, 8 lbs.; legume hay, 8 lbs.; oats, 5 lbs.
6. Oat or barley straw, chopped, 6 lbs.; legume hay, 10 lbs.; oats, 6 lbs.

Idle horses and mules, weight 1,200 lbs.

1. Grass hay, 17.5 lbs.; linseed meal or other high-protein supplement, 0.75 lbs.
2. Legume hay, 17 lbs.
3. Legume hay, 9 lbs.; grass hay, 9 lbs.
4. Corn or sorghum stover, 11 lbs.; legume hay, 8 lbs.
5. Corn or sorghum silage, 15 lbs.; oat or barley straw, 6 lbs.; legume hay, 7 lbs.
6. Oat or barley straw, 6 lbs.; legume hay, 12 lbs.

Brood mares nursing foals, but not at work, weight 1,200 lbs.

1. Alfalfa, soybean, or cowpea hay, 16 lbs.; corn or other grain, 6 lbs.
2. Red clover hay, 16 lbs.; oats or ground barley, 3 lbs.; corn, 3 lbs.
3. Mixed clover-and-timothy hay (containing 30% or more clover), 16 lbs.; oats, 6 lbs.
4. Timothy or other grass hay, 16 lbs.; oats, 3 lbs.; bran, 3 lbs.; linseed meal or other high-protein supplement, 1 lb.

Two views of an Australian big hitch. John Ashton of Kaleen, Australia, is plowing with 16 head (15 Clydesdales and one Percheron set in two 8 abreast spans) hitched via chain and pulley eveners to an 8 furrow stump joint disc plow (or plough as they say down under). For more information on this style evener system see Chapters 9, 11, & 12.

CHAPTER SIX
THE STABLE

Aden Freeman, of Ontario, Canada, at work harnessing his horses in single tie stalls. Photo by Kristi Gilman-Miller.

Comfortable quarters, conveniently arranged, are a help in keeping work horses healthy, in thrifty shape, and efficient as a power source. The barn or stable need not be expensive or fancy but it should provide ample room for the number of horses and mules kept, plus storage for plenty of feed and a separate room for harness. You may want to include, in the same structure, covered space for miscellaneous hitch gear or possibly even some wagons. It is a good idea not to include storage space for any flammable or combustible materials either in or near the stable. Blacksmiths, welding or mechanics shops should not be included in or near the stable nor should petroleum fuel tanks.

If you are building or rebuilding a structure to serve as a work horse stable, it is a good idea to consider certain working dynamics. For instance, with each stall filled with an animal and one or two animals harnessed, how difficult will it be for those working horses to be taken in and out of the barn? How are the passageways arranged? Are you expecting to have to lead a harnessed gelding close behind a tie stall that contains a pregnant mare or a stallion? Are the stalls set up so that some might be used, free access, from an outside paddock area by a convalescing animal or a stallion?

Another regular dynamic that needs to be considered in stable design is cleaning. Will you be able to clean the barn easily or will it require some special, time-consuming maneuvers? It is difficult for the

A sideview of the stabling interior of Aden Freeman's old style barn which features overhead hay storage. Photo by Kristi Gilman-Miller.

beginner to have any concept of convenience or inconvenience in stable design without some working experience with using horses daily. Some understanding might come from visiting a working farm that uses horses and paying attention to the time and physical requirements of stabling during the beginning of the day, at noon, and in the evening. One thing is certain: if the stable design, location and/or condition causes the teamster to take more time than is absolutely necessary in getting his animals to work, and then getting them settled after work, then the building is contributing to the inefficiency of horses as power. If the building works smoothly, it is adding to the efficiency of the horses as power.

There are many types of horse barns or stables varying widely in material and structure. In warmer southern climates barn buildings need not be so weather-tight and substantial, as is the case in harsher northern climates. The two-story gambrel-roofed building with hay storage overhead is an excellent structure for both its inherent insulating features and the ease of feeding.

The horse barn should be located on a comparatively high piece of ground so that the drainage will be good. A poorly drained, muddy yard is hard to keep clean, and the mud is likely to contribute to thrush (hoof rot) and other ailments. And, of course, the barn and yard should be located so as not to pollute water supplies.

Some farmers use large open sheds, feeding horses together at a free choice trough or manger. This setup may work for fattening idle animals, but it is not recommended for work horses. When horses or mules are at work, the time available for eating, especially at noon, is limited, and the timid animals may be crowded away from the trough so that they do not receive their share of the feed. This is why separate stalls with individual mangers are essential arrangements for the work horse.

Some say that box stalls are more comfortable than tie stalls. The author prefers tie stalls because they take up less barn space and because the horse that uses

Some people prefer box stalls and, if properly constructed, they do work well. However they require more time getting to the animals for regualr trips with teams to the field. Photo by Lynn Miller.

a tie stall MUST develop better barn manners than the animal that is shut in a box stall. It is possible to set up the barn with convertible stall space so that a box stall is easily used as a double stall or two tie stalls when necessary. This is the best plan when time, money and available space is a concern.

Whether or not you decide to use convertible stalls, your stable should include a bare minimum of one large box stall for use in the case of sick animals, foaling and some special instances that requires free confinement. The size of box stalls vary with minimum requirements of a work horse being at least ten feet wide either way and up to fourteen feet wide either way. Tie stalls should be an absolute minimum of four feet wide with five feet being a better width. Six feet wide is a waste and may allow the spoiled youngster to turn himself halfway around. The tie stall should be six to eight feet long (or deep) to the manger. Double tie stalls work best when 10 feet wide. The enclosed diagrams illustrates some stall variations with dimen-

sions. A well-tamped clay floor is considered to be the best for horse stalls (never use concrete or asphalt), and wood is a second choice. An expensive option is rubber or cork.

Windows should be placed so that at no time during the day must an animal look directly into the sunlight. If a horse or mule must look for hours into direct sunlight, it can cause permanent damage to the eyes. The easiest thing to do is put the windows high above the eyes of the horse. Horses rarely suffer from cold if kept dry and protected from drafts but they do require adequate ventilation. Make sure there is plenty of fresh air in the barn.

Feed (grain) boxes and mangers must be built of strong materials, just like the rest of the stall, as idle horses will tear them apart if otherwise. With tie or box stalls, the manger should run the entire length of the front (or one side) as this allows, in the case of box stalls, for the division into tie stalls and, in the case of tie stalls, is necessary to hold enough feed for the

These double tie stalls, which allow a team to stand together in each unit, are the author's preference. With these units, the horses can be fully rigged, unhooked from manger ties and backed out to go out for hitching. On the return, after unhitching, the teams may be ground driven right into their stalls. It is an easy matter to rig up removeable gate panels converting one of these stalls to a custom box affair to house foals when working mares.

Many of the old-style Gambrel-roof dairy barns were outfitted with tie stalls for work horses. Photo by Kristi Gilman-Miller.

animal. Page 105 illustrates design and dimension of mangers and feed boxes. This author prefers a manger with a false bottom up off the floor for two reasons. First, it will make for a shallower manger, reducing the amount of space in which a horse's nostrils might be trapped with dust. (If the bottom boards are spaced a sixteenth of an inch apart, the dust will fall through.) And second, it allows the manger to be higher so that anxious young stock or bored adults do not step into the manger.

When designing your barn, set up grain storage space so that it is impossible for a loose horse to get in to it, for if an animal should get into free choice grain it will eat until it gets colic and dies.

The harness room should be located very near the stalls of your working stock so that the harnessing chore is made easier. It is not a wise practice to keep harness hanging behind the horse because the manure and urine give off fumes that will contribute to the leather breaking down. (For an example of this, set an old, unwanted leather boot near or on a fresh manure pile for a couple of weeks and then test the leather.) In high moisture areas the teamster may want to protect his or her harness investment with a double-walled (maybe even

A roomy, single tie stall, well bedded with straw and ready for a horse. Tie stalls make for better, more cooperative horses because the animal must be led in, tied up, untied, backed up, accept confinement, stand in one place for a long period of time, asked to move over for grooming and harnessing, etc. All of these actions, as responses to commands, reinforce the training. Most of these actions are missing with a box stall arrangement.

insulated) tack room that can be kept shut. In the winter a light bulb left on will have a drying effect on the air in a small room. See the *HARNESSING* chapter for information about proper handling and storage of harness.

Keep in mind that the barn is advertisement to the public of what using horses means. A dirty, ill-kept barn with junk and debris everywhere is, of course, a hazard to the work horse, but it is also telling the public that a "teamster" is not taking care. Advertisement can go the other way as well. The author's favorite horse barn is on the Bear Paw Ranch of Gary Eagle where on the big swinging doors there is painted, "EVERY FARM NEEDS A TEAM."

Gary Eagle of Chesaw, Washington, had a handy log barn with the right advertising sign on its doors. The horse was his Belgian stallion during the 1970's. Photo by MaryLyn Eagle.

Tie Stall Plans

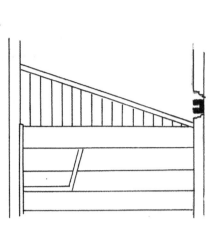

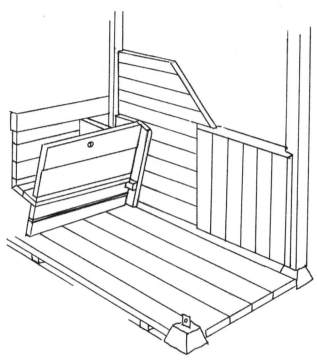

Above: One style of tie stall. The divider above showing position of manger and a cutaway illustrating cut channel in post designed to receive side boards.

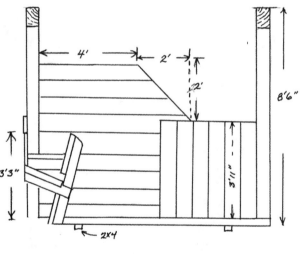

Above and Left: Another style of tie stall construction with dimensions.

The drawing below shows a possible position of this stall in a Gambrel-style barn.

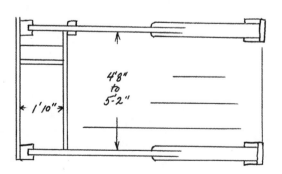

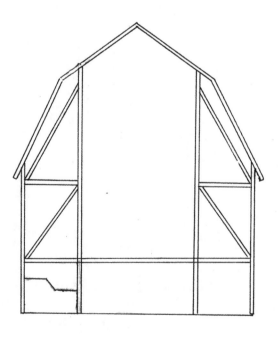

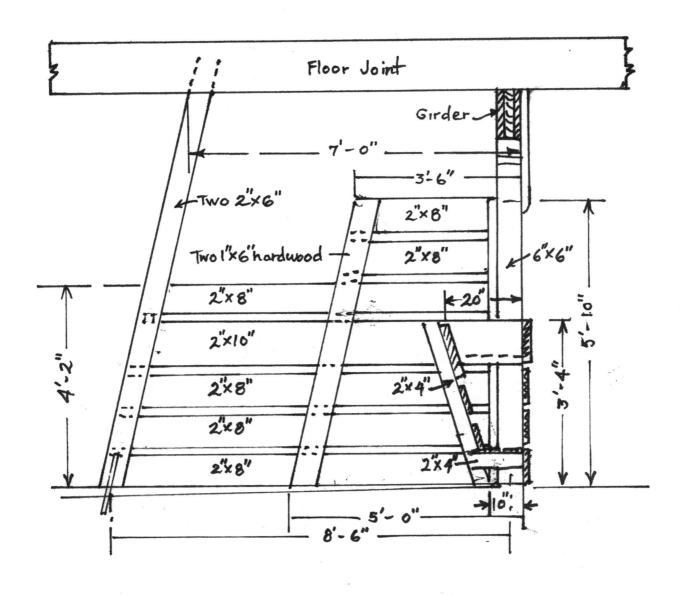

Floor Joint

Girder

7'-0"

3'-6"

Two 2"x6"

2"x8"

Two 1"x6" hardwood

2"x8"

6"x6"

2"x8"

20"

5'-10"

2"x10"

2"x8"

2"x4"

3'-4"

4'-2"

2"x8"

2"x8"

2"x4"

5'-0"

10"

8'-6"

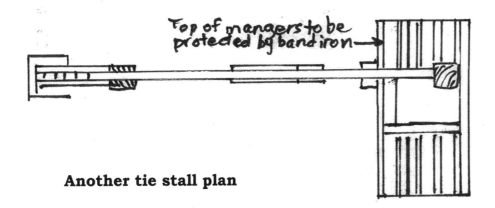

Top of mangers to be
protected by band iron

Another tie stall plan

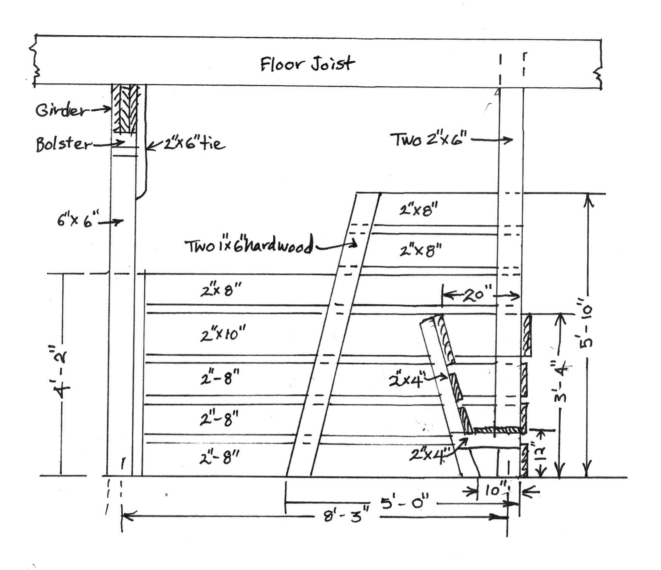

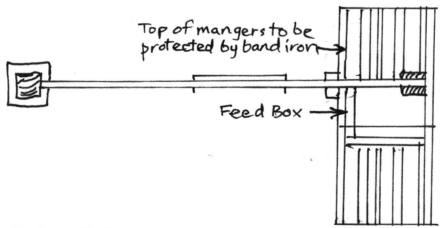

A fourth tie stall plan.

CONVERTIBLE DOUBLE (TEAM) TIE STALL

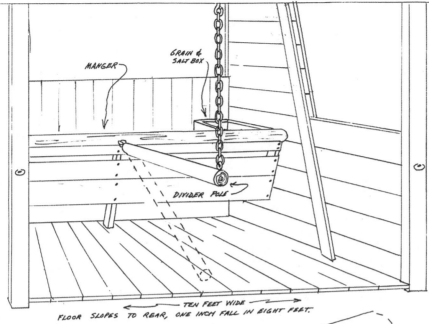

MANGER

GRAIN & SALT BOX

DIVIDER POLE

TEN FEET WIDE

FLOOR SLOPES TO REAR, ONE INCH FALL IN EIGHT FEET.

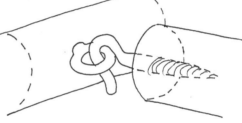

In this tie stall plan there is accomodation for a removable divider pole. By fastening two chains across the rear, from wall to pole end to wall, it is relatively easy to set the pole as rigid. This becomes an excellent training restraint for young horses, forcing them to stand straight ahead. The pole is easily removed and the open double stall becomes handy for the harnessed team to back out or drive in.

J.W. FISKE IRON WORKS, N.Y.

An extra fancy combination stabling setup from the turn of the century.

Above: The single tie stall is, on many farms, the work horse stabling of choice. In this picture we see ropes strung at the back to prevent the horses from backing up. Such a practise may only be helpful when alley space is limited and the farmer wants to keep the horses from backing up to kick at a passing horse. It has been my experience that, while the single stall provides security for the timid animals, it does seem to aggravate notions of territory. When I stable my horses together, as teams in double tie stalls, they frequently learn to get along with each other. And I have the distinct advantage of being able to drive them into the stall and back them out fully rigged. I also enjoy the added advantage of more physical space for harnessing. Look at the photo above and imagine entering the stalls with an armload of harness. It can be done, it is done everyday, there is a risk.

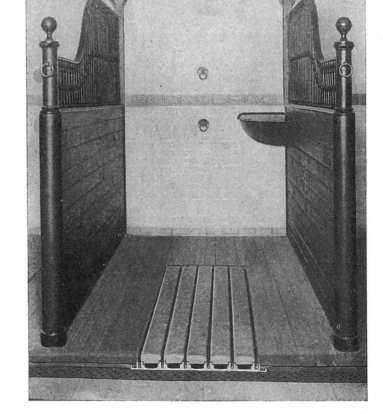

Right: An extra fancy store-bought tie stall setup. Note the drain channels in the floor.

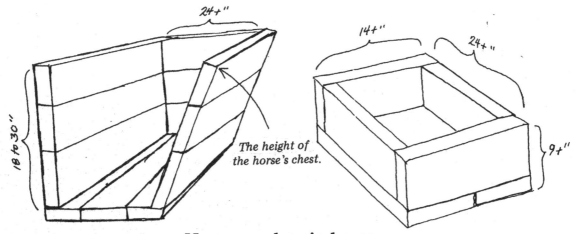

Manger and grain boxes

18 to 30"

24+"

24+"

14+"

9+"

The height of the horse's chest.

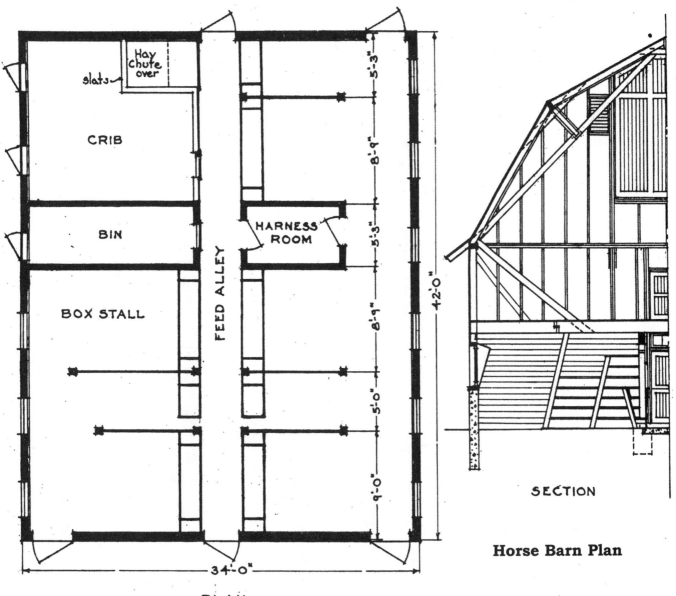

slats

Hay chute over

CRIB

BIN

BOX STALL

FEED ALLEY

HARNESS ROOM

5'-3"

8'-9"

5'-3"

8'-9"

5'-0"

9'-0"

42'-0"

34'-0"

PLAN

SECTION

Horse Barn Plan

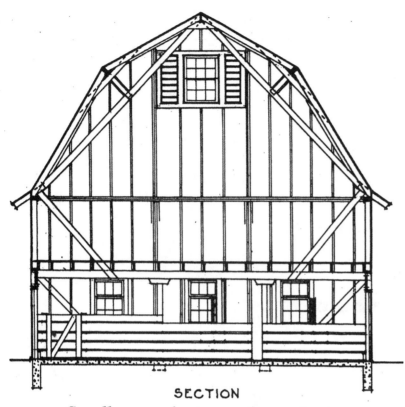

SECTION

Small general purpose barn plan

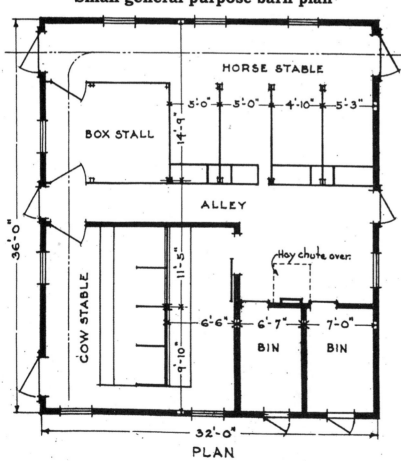

HORSE STABLE

5'-0"   5'-0"   4'-10"   5'-3"

BOX STALL

14'-9"

ALLEY

36'-0"

Hay chute over.

11'-5"

COW STABLE

6'-6"   6'-7"   7'-0"

9'-10"

BIN   BIN

32'-0"

PLAN

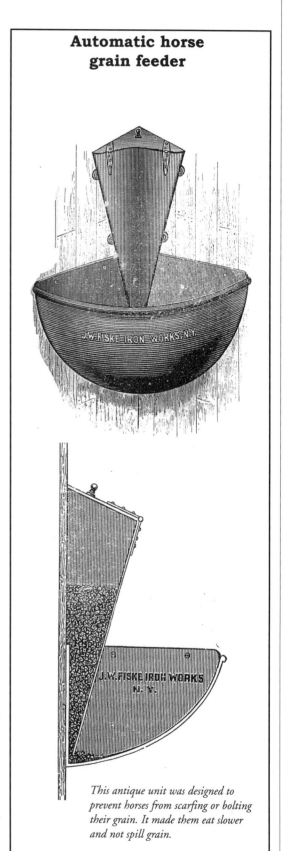

Automatic horse grain feeder

J.W. FISKE IRON WORKS NY

J.W. FISKE IRON WORKS N.Y.

This antique unit was designed to prevent horses from scarfing or bolting their grain. It made them eat slower and not spill grain.

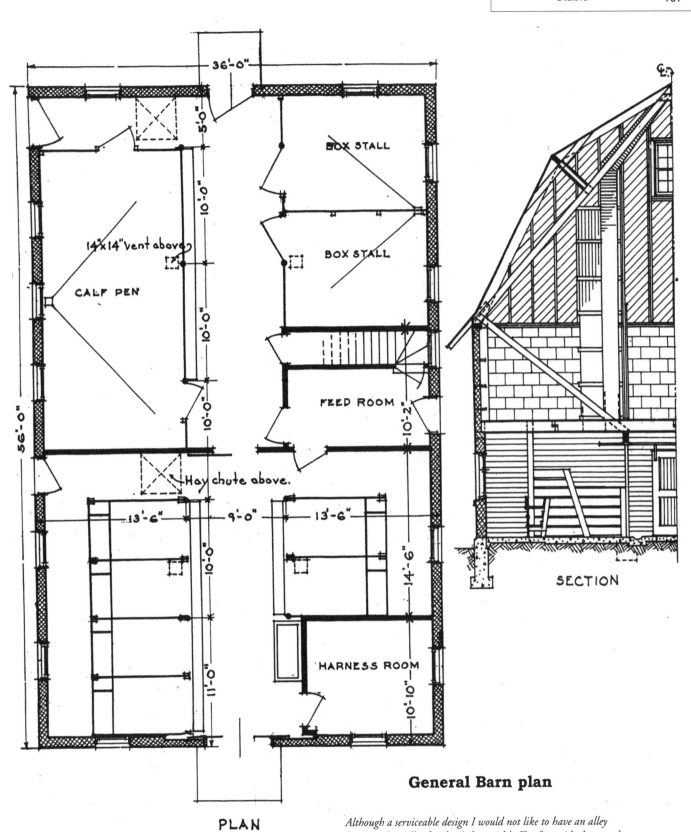

36'-0"

5'-0"

10'-0"

BOX STALL

14x14"Vent above

10'-0"

BOX STALL

CALF PEN

10'-0"

56'-0"

FEED ROOM

10'-2"

Hay chute above.

13'-6"   9'-0"   13'-6"

10'-0"

14'-6"

11'-0"

HARNESS ROOM

10'-10"

General Barn plan

PLAN

Although a serviceable design I would not like to have an alley between the stalls of only 9' clear width. Ten foot wide doors and a clear alley width of 12 foot is preferred.

SECTION

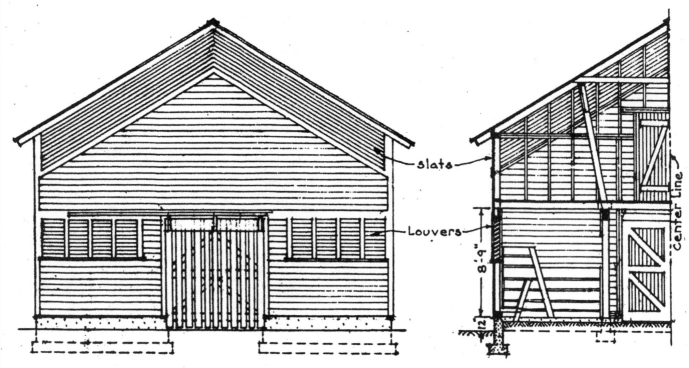

SOUTH ELEVATION

slats

Louvers

8'-9"

Center Line

SECTION OF ONE HALF OF BUILDING

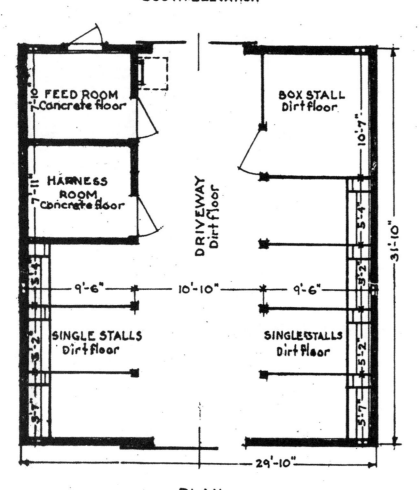

FEED ROOM
Concrete floor

HARNESS ROOM
Concrete floor

BOX STALL
Dirt floor

10'-7"

5'-4"

3'-2"

DRIVEWAY
Dirt floor

9'-6"   10'-10"   9'-6"

31'-10"

5'-4"

5'-2"

SINGLE STALLS
Dirt floor

SINGLE STALLS
Dirt floor

5'-2"

5'-7"

5'-7"

29'-10"

PLAN

Horse barn plan

This is an excellent floor plan for a working horse barn easily modified to additional box stall(s) or double tie stall(s). The only fault I can find with this plan would be the width of the alley. I'd like to see at least 12 feet clear. I would also opt for wood stall floors.

CHAPTER SEVEN
HARNESS

This chapter is divided into two sections. In the first part, with minor revisions, we have left the first edition's simplicity intact. The second part, under the heading of *additional information*, we have added new innovations and greater complexities. Beginners may choose to skip or skim the additional information if they sense a detail overload.

In order to take maximum advantage of the horse's inherent capacity for pulling (or draft), man, through the ages, has designed rigging, the specific purposes of which were:

• First to hang, fasten or balance (otherwise anchor) ropes, chains or leather straps to a point on the shoulder of the animal, and in this manner result in a system for pulling.

• Second, to outfit the head and mouth in such a manner that a person could control and guide the animal.

• Third, if using a wheeled vehicle, provide some manner in which the vehicle could be backed.

The history of the development of harness is fascinating and includes a rich variety of different designs and solutions. There is not room in this text to cover the history of harness. Instead we are limited to a discussion of modern-day western-type harness. (Under additional information we will discuss harness built of synthetic materials.) Hopefully this examination will provide some basic understanding of how and why harness works.

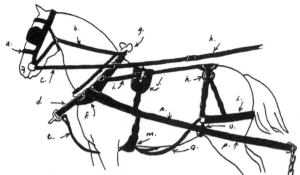

A. Western brichen harness (parts are identified on next page)

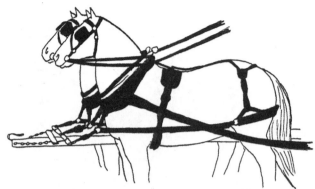

B. Boston side-backer harness (see information on page 156).

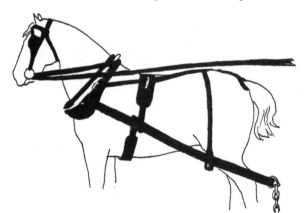

C. Market tug harness. (No back pad)

D. Breast collar buggy harness

E. Plow, or cruper, harness. (No brichen)

F. Collar buggy harness

styles of harness

Figure A on this page is a diagram of a standard design farm team harness referred to in this text as a western *brichen harness*. The important parts are identified. *Figure C* illustrates another design of farm team harness referred to in this text as a *market tug* harness. The notable differences between the western brichen harness and the market tug design are the absence of the back pad (or saddle) and the two piece tug. *Figure E* is a diagram of a brichen-less harness referred to in this text as a *plow* harness. *Figure B* is a sketch of a *side-backer harness* which is more complex and quite popular in eastern regions. *Figs. D & F* are examples of light buggy harness, one breast strap and the other collar-style. These harness styles made be constructed with either leather or synthetic materials such as BioThane or a mix of materials. They will function the same regardless of materials used. An understanding of how and why these basic styles differ in function and comfort may come from a discussion of the individual parts.

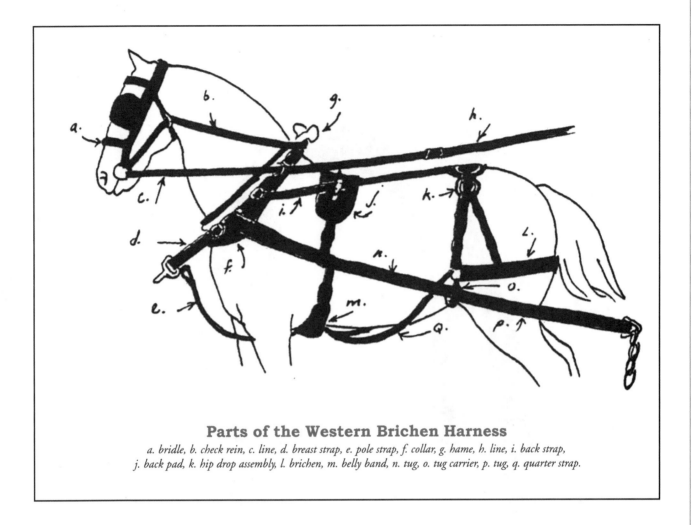

Parts of the Western Brichen Harness

a. bridle, b. check rein, c. line, d. breast strap, e. pole strap, f. collar, g. hame, h. line, i. back strap, j. back pad, k. hip drop assembly, l. brichen, m. belly band, n. tug, o. tug carrier, p. tug, q. quarter strap.

parts of the harness

The Collar

Pages 112- 116 illustrate the different popular shapes of collars. Reasons for the different shapes include:

• Horses and mules have different shaped necks with mules commonly being more flat-sided.

• Stallions have thick necks with big crests and require larger-wider collars in order that they set back against the shoulder in the required manner.

• Some animals, because of the shape of the neck, will require additional room at the bottom of the collar to free the windpipe.

• The shape, both in terms of condition and position of the shoulder, can affect how well a collar works, for this reason the width of the draft (or widest part of the collar) could make a difference.

• Personal preferences of different teamsters have an effect on some of the aspects of collar design.

NOTE: When storing collars in a hanging position, always turn upside down. Hanging right side up (against the collar-cap) will result in a slow distortion of the collar's shape that will be difficult, if not impossible, to correct. If a warped or twisted collar is fitted to a green horse, it could possibly result in some damage to the horse's shoulder and/or the horse's attitude towards work.

The illustration on page 117 shows the proper place of the collar around the horse's neck. (See HARNESSING chapter for correct way to put the collar on.) The collar provides a place from which to fasten the hames and tugs and on which to attach the remainder of the harness. Proper fit and condition of the collar is critical to the horse's well-being and capacity for work. A well-crafted collar has no value if it does not fit. Vice versa, a poorly constructed or dilapidated collar may fit perfectly and not work properly.

If a collar is TOO SHORT, TOO LONG,

Mule Collar

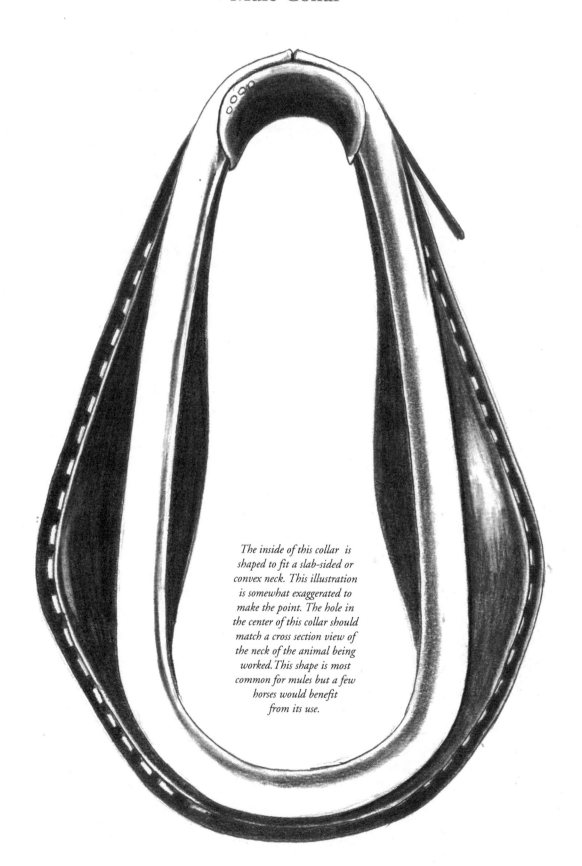

The inside of this collar is shaped to fit a slab-sided or convex neck. This illustration is somewhat exaggerated to make the point. The hole in the center of this collar should match a cross section view of the neck of the animal being worked. This shape is most common for mules but a few horses would benefit from its use.

Half-sweeney Collar

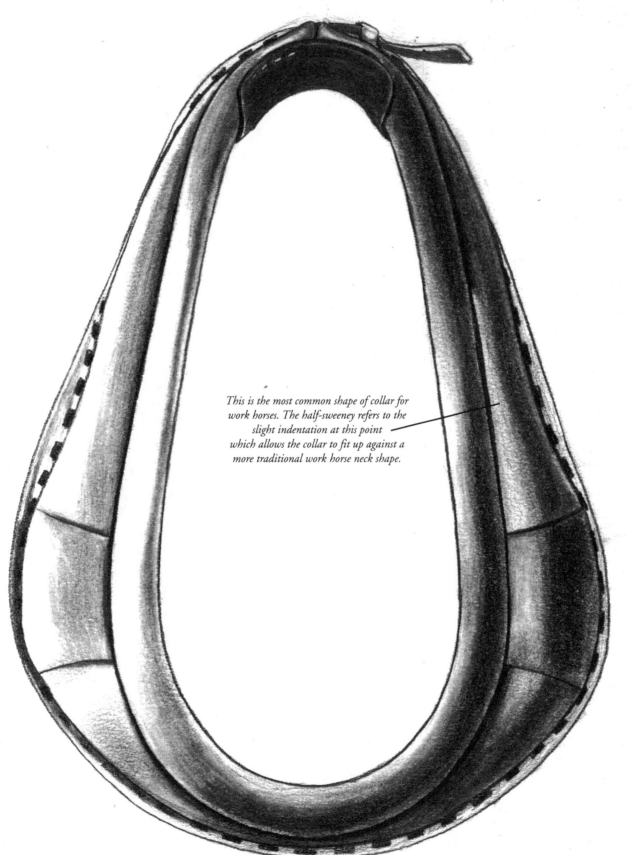

This is the most common shape of collar for work horses. The half-sweeney refers to the slight indentation at this point which allows the collar to fit up against a more traditional work horse neck shape.

Half-sweeney Pipe Throat Collar

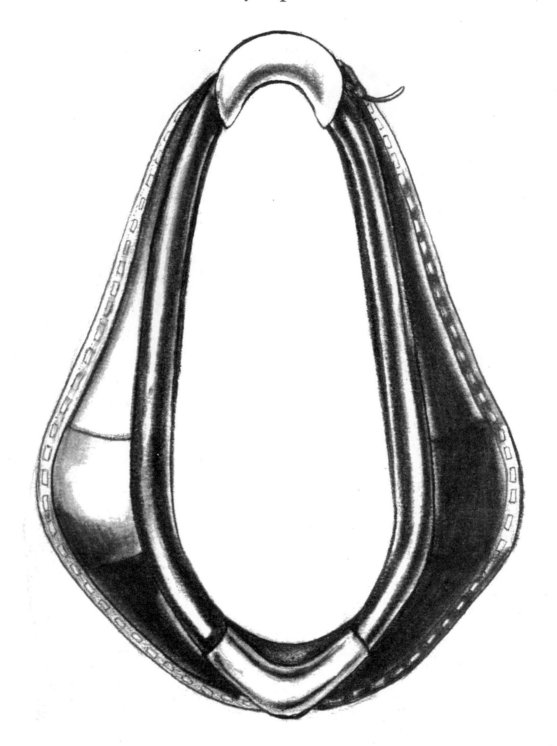

It is our impression that this style of collar in no longer being made. That is too bad for horses and teamsters. With the added room, made available by the bent pipe shape (which also reinforces the throat of the collar structure), it is far easier to get a good fit on the horse than with the conventional throated collar styles. You want a good snug fit which doesn't cut off the animal's wind. Perhaps someone will come along in the near future to build this style of collar.

Full Face Collar

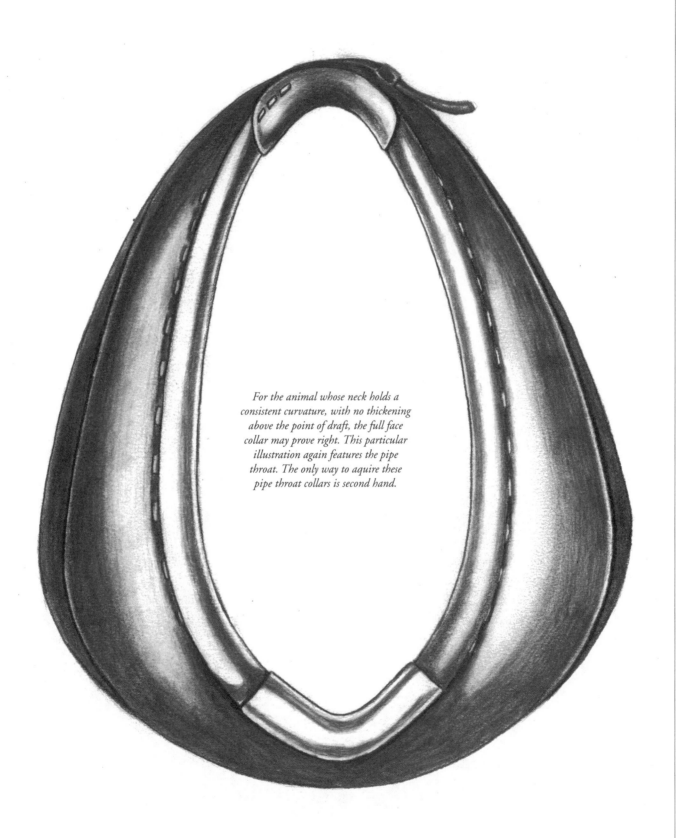

For the animal whose neck holds a consistent curvature, with no thickening above the point of draft, the full face collar may prove right. This particular illustration again features the pipe throat. The only way to aquire these pipe throat collars is second hand.

Full Sweeney Collar

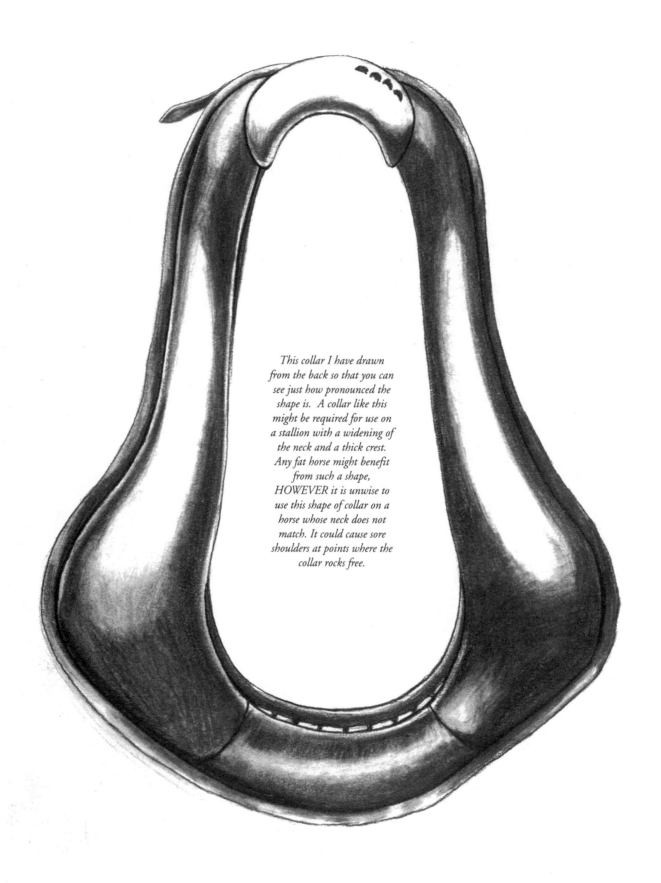

This collar I have drawn from the back so that you can see just how pronounced the shape is. A collar like this might be required for use on a stallion with a widening of the neck and a thick crest. Any fat horse might benefit from such a shape, HOWEVER it is unwise to use this shape of collar on a horse whose neck does not match. It could cause sore shoulders at points where the collar rocks free.

The fit of the collar is checked at the windpipe.

horse is worked. With the possible exception of a stallion, excessive weight (with resultant crest) will be reduced and probably disappear when the horse is toughened-in or conditioned. Even the well-fed horse in show-type condition (or flesh), if in good working muscle tone, will not have an excessively heavy neck.

Equally important to the prevention of sore shoulders on the horse is the correct position of the hames so that the point of draft (or point at which the tug attaches to the hame), is where it should be (see pages 32, 34). The hames must be the proper length so that the curvature conforms to the groove (or hame seat) of the collar. The hames are measured, as illustrated on page 118, from the bottom to the uppermost hame strap fastening point. Even with hames of proper length it is possible to have them incorrectly adjusted on the collar. Common error is to either have them too high on the collar or offset with one side high and one side low. This causes problems for the animal. The hames should center near the widest point of the collar (the draft) and the collar draft should be on the point of the horse's shoulder.

measuring and fitting the collar

Page 118 illustrates the procedure for measuring a collar. This measures 20 inches on the ruler. For the size of draft draw a tape measure around the collar at its widest point as illustrated.

It is difficult to accurately measure a horse's neck for a collar without fitting. In other words, there are so many variables involved in the size and shape of a horse's neck that the only accurate and easy way to size the neck is to use several collars and put them on one at a time until fitting is found. In the heyday of the harness horse, collar makers had sliding shape-conforming measuring devices which were used.

The horses neck is measured from just ahead of the withers down along the neck to the windpipe, in a straight line. This measurement is not taken on the curve but straight as shown on page 118. To order collars it would be helpful to be able to get a width measurement as well. Our friend, John Erskine, shared

TOO WIDE or TOO NARROW it will cause both the horse and the teamster misery. The drawing above illustrates the proper fit of the collar which requires enough room between the horse's windpipe and the bottom of the collar to allow the fingers of a flat hand to pass freely between the windpipe and the inside throat of the collar when it is seated against the shoulder. Also, the sides of the collar should fit cleanly, not tight, with perhaps a finger's width clearance at best. The collar should seat flatly against the shoulder and not rock on a wide spot in the neck. To this end, the full-face collar is for a flat-sided neck; the half-sweeney collar for a slightly full neck and a full sweeney collar is for a thick, full neck.

NOTE: When fitting an overweight horse to a collar, keep in mind that the neck will lose size as the an ingenious idea for arriving at a width measurement for the horse's neck. He holds two carpenter's framing squares, with the short legs over the top of one another and the long legs on each side, forming a U. He slides this up from under the neck and adjusts to measure at the widest point.

hames

The hame, page 119, is usually of steel or steel reinforced wood construction. It is rib-like in design and two are used with each harness. They are strapped top and bottom, tightly, into the outside groove of the collar. The tugs attach to the bolt (or hook) assembly (at point of draft), the breast strap to the bottom ring, the back strap to the second ring and the driving line passes through the top ring. The brass or nickel-plated balls found atop the hames are purely ornamental. On some hames, small rings may be found just below

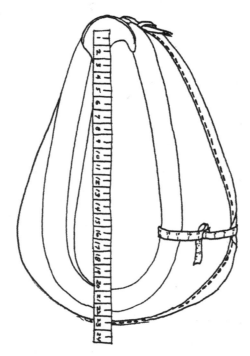

the hame ball. These rings are used to snap spreader straps into.

It is customary to leave all the working straps of the harness attached to the hames when stored. See HARNESSING chapter.

It is normal to have three or at least two adjustable top hame strap positions and for a pair of hames to be adjustable up to 3 inches in length.

The length of hames, of course, corresponds to the size of the collar. In other words, when referring to 22" hames we are talking about the size of the hame and not its overall length. To determine the size of a hame, measure, see below, from the bottom of the hame straight to the point at which the top hame strap fastens. Since this point on the hame is adjustable, you'll find that it will commonly measure to within a two inch range, for example 20"-22" or 26"-28". It is necessary to match the size of hames to the size of the

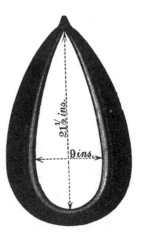

HAMES & HAME STRAPS

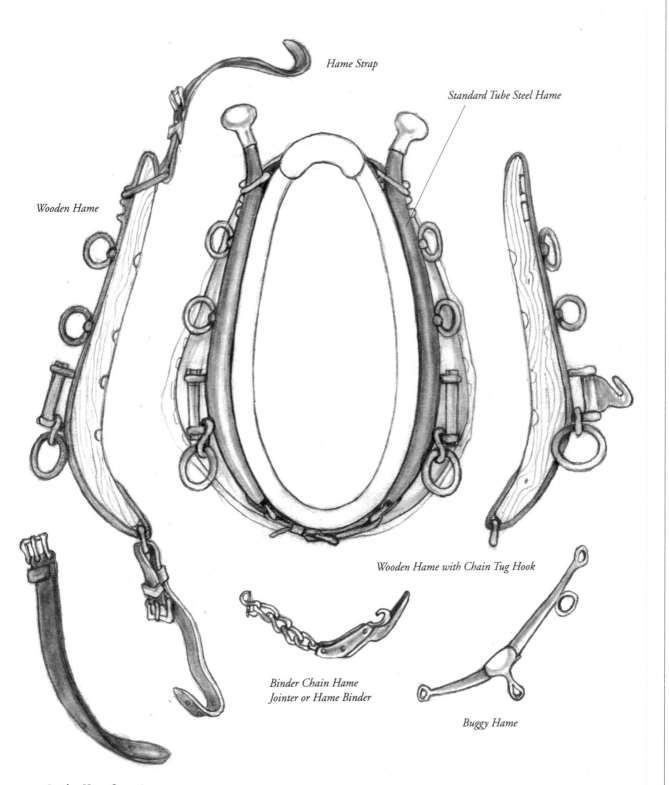

Hame Strap

Standard Tube Steel Hame

Wooden Hame

Wooden Hame with Chain Tug Hook

Binder Chain Hame
Jointer or Hame Binder

Buggy Hame

Leather Hame Strap

collar. See the section, this chapter, under COLLARS which refers to the placement and adjustment of hames

hame straps

The hames are secured top and bottom into the groove of the collar by the use of hame straps, see page 119. These short stout straps are commonly made of leather, sometimes nylon and less frequently bindered chain. In the case of leather and nylon, the strap normally passes through the right side hame (one strap top and one strap bottom) first and then the left side so that cinching up the straps can be done easily from the left side of the animal.

The strength and proper adjustment of hame straps are critical to proper performance and safety. If a strap should break while pulling, dangerous injury may well result. If a too loose strap should cause the hames to pull forward out of the collar groove, the animal might choke severely. A strong, properly and tightly adjusted hame strap is important.

[In pulling contests or heavy logging it is not uncommon to see hame straps that are stitched two-ply thick. Also, second hame straps will be used top and bottom. On top a strap will pass around the hame and twist around the other hame. From the top this would look like a figure eight. On the bottom a longer strap will pass through the breast strap ring (below the tug) and on top of the hame strap over to the same ring on the other side. Then a regular length hame strap (or shorter) would pass completely around the collar at the windpipe, and, of course, over both hame straps.]

Cinched down, this strap serves to hold the hame strap assembly from slipping up to the windpipe. To appreciate how critical the hame strap is to pulling performance and all work in harness, one need only inspect the care that professional pullers take with this part of their harness.

The hame strap, because of its handy size, is the perfect extra part to keep on hand for alterations or quick replacements or repair. The author recommends that the teamster keep one extra hame strap for each

harness regularly used.

tugs or traces

Tugs, or traces, as they are frequently called, attach by a steel pin, bolt or hook, to the hames. They are attached directly or indirectly, at the opposite end to whatever is to be pulled. It is critical that they be strong and pliable. Tugs today are commonly contructed of multi-ply sewn leather, BioThane, nylon webbing, combinations of leather and nylon, and less frequently of rope or chains. There are many different designs of tugs, sometimes varying in strength, length and hookup, depending on the job they are to per-form. The leather farm tug is commonly fixed with a chain at the end by which it is attached to the single trees. In some cases, more common with logging or pulling contest harness, the tugs have metal hooks on the ends and an adjust-able butt chain with large rings on both ends (one for each tug), fastened to the singletree. (See chapter RIGGING THE HITCH.)

When the horse is pulling, he is actually pushing with his shoulder against the collar. That pushing action is transferred through the tugs into a draft or pull. If a tug should break while the animal is pulling, a fall would result that could cause damage. Since the animal has two tugs, one on each side, it is important that there be no mechanical reason preventing the horse from moving cleanly and freely between the tugs, especially if they are chain or rope.

back pads

The back pad, or saddle as it is sometimes called (see page 122), seats just back of the horse's withers as far forward as the animal's conformation comfortably allows. It properly carries and evenly displaces any downward pressure or weight that might come from either tugs (on the draft angle) or shafts (such as with a single horse hitched to a cart). In most team harness the back pad serves primarily to organize the girth position of the belly band and to support the traces through the billets. Since it does not, in common western use, carry much weight, its size is lighter than

TUGS & BELLY BANDS

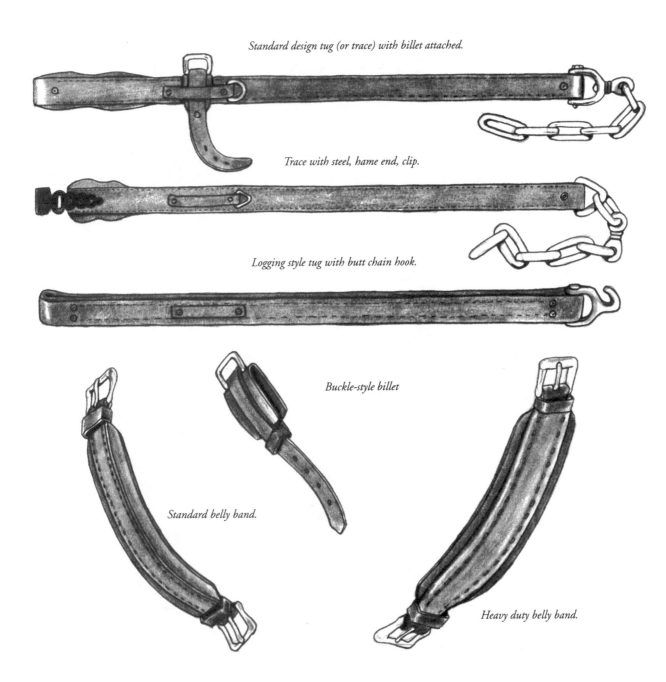

Standard design tug (or trace) with billet attached.

Trace with steel, hame end, clip.

Logging style tug with butt chain hook.

Buckle-style billet

Standard belly band.

Heavy duty belly band.

Tugs or traces may be made from single and multi-ply leather, coated or uncoated nylon
webbing, chains with leather sleeves at prominent wear points, or on rare occasions rope.

BACK PADS, BRICHENS, HIP DROPS & CRUPER

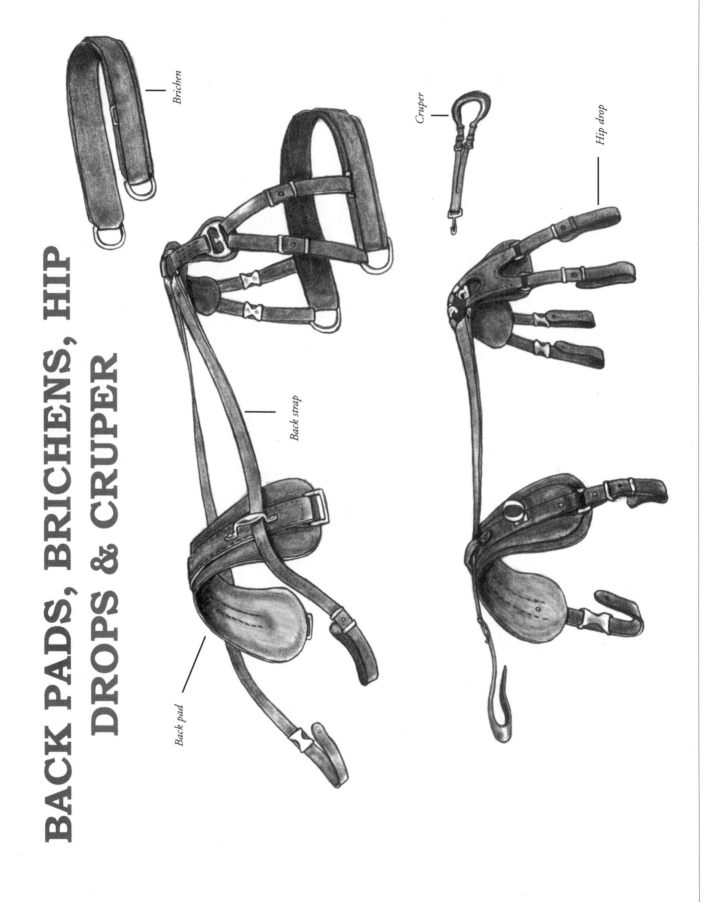

Brichen

Cruper

Hip drop

Back strap

Back pad

BREAST STRAPS, POLE STRAPS & QUARTER STRAPS

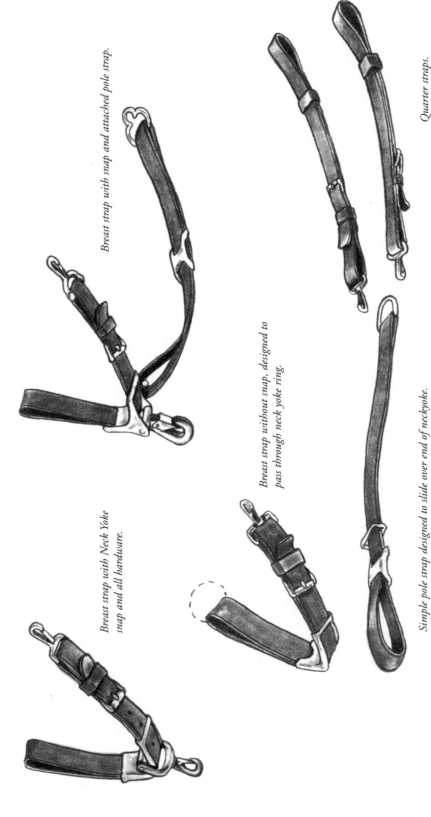

Breast strap with snap and attached pole strap.

Quarter straps.

Breast strap with Neck Yoke snap and all hardware.

Breast strap without snap, designed to pass through neck yoke ring.

Simple pole strap designed to slide over end of neckyoke.

Pole strap with swiveling neck yoke snap.

A brand new leather harness, no hames, built by retired harness-maker Bill Burckle of Kentucky.

European harness where horses are commonly used single and a greater weight is transferred through the shafts to the horse's back. Indeed, in some styles of western harness, such as the market tug harness (see page 110, fig. C), the back pad is done away with altogether.

Some styles of back pad will have billet straps sewn into either end for fastening around the tugs. More commonly the back pad will have "D" rings in either end through which back pad billets may be fastened.

billets

Billets are short heavy connecting straps which go around the tug or fasten into the back pad and have a free end to enter a buckle (see page 121).

breast straps

The breast strap is a wide, long, strong strap which fastens from the lower hame rings on both sides of the hames, see page 123. There are a wide variety of different setups for the breast strap as it works with or without the pole strap in its function with the neck yoke. The author prefers the all-in-one assembly where the pole strap is joined to the breast strap by a bolt and roller assembly snap. Page 123 shows this and three other variations which include: no neck yoke snap on either pole strap or breast strap; snaps on both; a snap on just the breast strap.

The breast strap functions in an important capacity as part of the backing and braking system of the harness. In some regions, such as eastern Canada,

the breast strap works alone as the braking and backing system. The author cannot recommend such a practice where there are heavy loads to draw over steep terrain.

Another related function of the breast strap is, on rare occasion, to adjust the height of the end of the tongue, such as with a mower.

The breast strap is commonly looped through the right side bottom hame ring. It is then fastened by heavy snap to the left side bottom hame ring. Sometimes you will find a breast strap with snaps on both sides, usually in an all-in-one assembly. The breast strap must be strong because, if it should break at a critical moment, the load behind the team could come up on their back legs and cause trouble.

pole straps

The pole strap works with the breast strap and quarter straps to tie the collar and brichen together to function as the complete braking and backing system for the harness. This heavy strap usually has a heavy "D" ring in the back end and either a snap or loop in the front end (see BREAST STRAP description above). There are variations in design of the pole strap and those differences often need to match the design of the breast strap, see page 123. On the front end, the pole strap fastens in some manner to the neck yoke. On the back end, the pole strap is snapped into by the two quarter straps. In western style harness, it is most common for the pole strap to pass over the belly band loosely. In some regions, the pole strap will have a sewn loop in the back end that allows the belly band to pass through. Care should be taken that this style harness is properly adjusted so that the pole strap does not pull the belly band forward into the back of the horse's front legs when either backing or braking a load. Adjusting the length of the quarter straps would be the most direct way to prevent this.

The pole strap needs to be strong. Should it break at the wrong time, the load behind the horses could come up on them.

back straps

The back straps on a western style brichen harness run from the hip drop assembly forward on both sides of the animal and fasten into the hame ring just above the tug. In most cases these straps pass through loops on the back pad and are buckled or conwayed into the hame rings. In some instances, the back straps have snaps in the front ends for fastening to the hame rings (usually in a detachable brichen setup). In eastern Canada, turrets are used on the back pads for line guides so there is no room or place for the back straps to pass through back pad loops. What is commonly done in this case is to just allow the straps to run free over the back pad to the hame ring fastening point. It seems to work just fine.

The adjustment of the length of the back straps will affect the position of the hip drop assembly on the animal, and to some extent the position of the brichen. The back straps are normally adjusted by a sliding conway buckle (see HARDWARE illustrations).

hip drop assembly

The hip drop assembly (see page 122) is a set of connecting straps which serve to hold the brichen in its proper place. From center, top of the hip, hip drop straps, usually two to a side, sometimes three, run down to fixed positions along the brichen. The straps are adjustable in length to allow for correct height and position of the brichen. The entire assembly must be properly adjusted to seat and hang from the center of the top of the hips (adjust back straps) before the brichen can be properly adjusted for height and angle. The harness must be pulled back with the animal well into the collar before an accurate adjustment can be made. If the hip drop assembly is too long in the back strap adjustment, the brichen will hang low and not work in braking and backing. If it is low enough, it could cause the animal to refuse to back. (See adjusting the harness, chapter eight.)

brichen

The brichen is a heavy multi-ply leather (or synthetic) strap which wraps around the horse's backside under the tail. This strap is held in position by the hip drops. The brichen must be in the proper position if it is to function right. The strap should naturally hang level and work with the quarter straps as they pass under the belly, see page 122. The brichen should be loose when the animal is pulling a load. It should be tight when backing or when restraining or braking a load. The brichen works through its connection with the quarter straps to the pole strap and breast strap to the neck yoke. This forms the backing and braking skeletal structure for the harness.

quarter straps

There are two quarter straps to each harness. These straps loop and buckle through the "D" rings on the ends of the brichen. The front end of each quarter strap has a heavy-duty snap which fastens to the back end of the pole strap. If the quarter straps are adjusted too tight, they will cause the horse some discomfort by rubbing when the team is hitched to a tongue setup. If the straps are too long, they make the use of the brichen in braking and backing ineffective and a horse might easily step over the quarter strap and break the harness. Even when hitched and pulling there should be some slack in the quarter straps (about the width of two hands flat), but not too much. Only when there is a backing and braking action should the quarter straps come tight against the belly. The quarter straps should be strong.

crupers

The cruper is a strap which fastens, not too tight, around the animal's tail. In a brichenless plow harness, a cruper is used to hold the harness in place on the animal, so that the rear portion of the rigging doesn't slide off to one side or the other. Even in some brichen harness the cruper might be used for the same purpose. If there is an excessive pull against the cruper, it may well cause the animal discomfort and account for some unusual reactions.

Some harness is designed to have a check rein run through its connections to the cruper. There are new harness designs which have the cruper run through connections to the front ring ends of the brichen. As a design innovation, this appears to be an improvement for the sake of the horse.

Care should be taken when putting the cruper on any animal for the first time. (Page 122 illustrates the cruper.)

trace carriers

The trace carriers (sometimes called mud straps) are self-explanatory. They fasten to the "D" rings on the ends of the brichen (below the quarter straps) and serve to keep the traces, or tugs, up and out of the way.

In a cruper harness it is not uncommon to see trace carriers on longer straps fastening on the top center of the hip to the back straps. Certainly in many instances they are not used at all. In the southern U.S. where heat and economics dictate thrift, it is most

Work
Bridles

Work Bridles

common to see cultivating harness without trace carriers.

Though they may not be critical to the harness function, it is important to note that trace carriers set too high will pull up on the tug creating a downward push through the hip straps, when pulling. This may cause the horse discomfort at least and may be the beginning of creating a balky horse (or one which doesn't want to pull). Please see the troubleshooting section of the next chapter.

bridles

Page 126 illustrates two common variations in work bridles. There are, of course, many more designs. The bridle is a vital key in the teamsters' ability to communicate with the animal. This apparatus fits comfortably over the horse's ears and down the sides of the head to hold a bit in the mouth, preferably with slight or no tension, but certainly no slack. A throat latch passes under and is fastened, usually on the left side, with just enough room for a finger or two to pass between the throat and the strap. The bit is held in place by two bit straps, one on each side, which allow for some adjustment of position in the mouth. These bit straps are critical, as should one break or come undone you will have NO control over the animal. In a used harness, the bit straps are one item that the author recommends be replaced with new parts. Usually the bridle can also be adjusted for overall length at the top sides, over the blinders, or top center. In a bridle with blinders on, it is important that length be adjusted so that the animal's eyes come on center of the blinders. If the animal can see over or under the blinder, it will prove worse than ineffective.

As for the arguments for and against blinders, this author chooses to simply refer to the recognized purpose of the blinder. By restricting the horse's vision to the path it must follow, the teamster limits the chance that an unexpected sight might spook the animal. The animal that works comfortably without

blinders may react suspiciously to having to wear them. The same is true in reverse. Personal preference will play a big part in the experienced teamsters' decision about blinders.

Most bridles are set up with check reins. The check rein serves to hold the horse's head up in its natural standing position. It restricts the animal from putting its head down for grazing or rubbing. The check rein fastens into the bit, passes up and through a hardware ring (called a combination swivel), fastens to the throat latch, and then back along the neck, over the withers and forward again in the other side through the same pattern to the opposite bit ring. There are a variety of methods for fastening the check rein back. Some of the most common include hooking into some part of the back pad, snapping into a strap that comes forward from the hip drop assembly or simply adjusting the length of the check rein to allow it to work properly if hung over one or both of the hames.

In any event, the important thing here is not to overdo the check rein, and force the animal to hold its head in an unnatural position. To do so will cause you problems later, not the least of which will be the loss of confidence from your animal.

It is not uncommon to see animals being worked with no check rein. A skilled teamster will be able to identify how one animal, or one job, performs better with or without the check rein.

After a bridle has been fitted to a particular

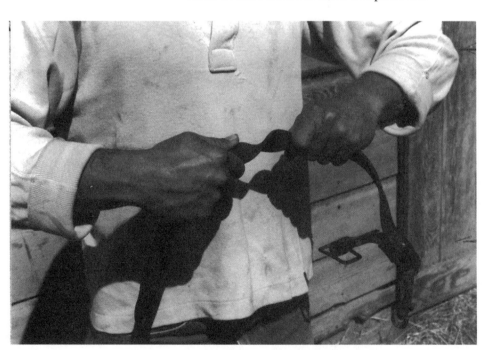

To test an old leather strap, twist it hard, and then harder.
If it is brittle or raggy it will break or tear.

LINES

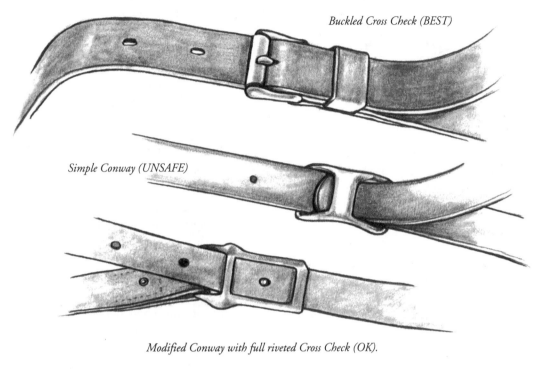

Buckled Cross Check (BEST)

Simple Conway (UNSAFE)

Modified Conway with full riveted Cross Check (OK).

Conway Cross Check

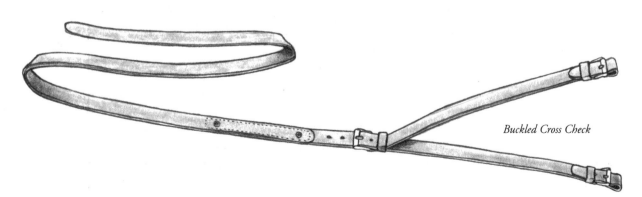

Buckled Cross Check

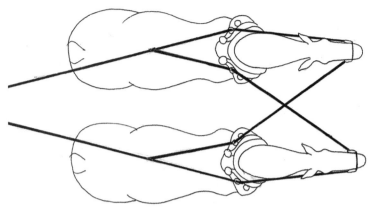

Showing the setup for team lines.

animal, it would be a good idea to keep it for that horse. In other words, each horse should have its own bridle, if not harness, as it will stretch and bend to shape like a shoe to your foot.

lines

Driving lines differ substantially depending on how many animals are being driven, in what configuration, and at what job. One horse is normally driven with two single lines with either snaps or buckles on the ends. A team of horses is driven with two checked team lines, see above. For the various differences, see the chapter, this text, entitled RIGGING THE HITCH and the illustrations of driving lines.

If you are working with used harness, make doubly sure that your lines are strong and pliable and adequately stitched or riveted. One way to check the tensile strength of line leather (or any old leather) is to twist a short section of the line severely and pull on it. The leather should not tear or even look suspicious. Also check the leather closely under any hardware, especially near the bit snap or buckle. And finally, check the splices to see if the stitching is beginning to rot or the splice is coming apart or any rivets are badly corroded. There might be "a lot of good leather" in a used line but these other points may be bad enough to guarantee disaster. A used line can be repaired and strengthened but until you know what you're doing, it is best done by a competent harness shop.

Unfortunate experience, namely a needless runaway, has caused this author to have strong and definite preferences in the construction design of team lines. Page 129 illustrates the differences and preferences.

A friend of the author and a true master teamster, the late Ray Drongesen, preferred not to use

lines with snaps for the bits. Instead, Ray used buckles in the end giving him just a little extra security. A horse, when standing at rest in harness, might play with the line, mouthing the snap and unsnapping the line. Then when the unsuspecting teamster gives the command to go, there is a surprise. This doesn't happen with buckled lines.

There is an altogether different and fairly uncommon connector we call a twister wire snap. Doug Hammill likes these and believes they are free of the problems common to most snaps.

Twisted Wire Snap

As illustrated in various places throughout this text, the basic setup for team lines is as follows: The main line length passes along the outside of the horse (and team) through the top hame ring and fastens below any other strap, to the bit ring or shank (snaps face out). The cross check passes over the same horse's back through the top hame ring and across to the teammate's bit ring. The same procedure is followed in reverse for the other line. There are some variations of setup which might include spreaders and/or center line drops, but the basic principle and the functioning dynamic remain the same.

The length of the lines will vary considerably with the job to be done or how far away from the horses the teamster must stand (or if using ponies). Customarily, the length is between 18 and 20 feet with some cases calling for lengths from 15 to 25 feet.

bits

There are thousands of designs of bits invented through the ages reflecting the human's ever-changing attitude about the best relationship and performance to be expected from (or with) the horse. Some bits are incredibly cruel to the horse's mouth and/or jaw. Others are so benign as to be almost ineffectual as they do not allow for a range of various sensitive pressures to the horse's mouth. Thank goodness there are some bits that combine gentleness and flex so as to provide the teamster with the fullest range of communication through the hands to the animal's head and mouth.

If you think of the bit as simply an iron tool in the horse's mouth which will stop the animal if pressure is applied, please read the chapter in this text entitled ATTITUDE AND APPROACH. The author's prefer-

Bits

Above and below: Two slight variations on the classic snaffle bit. This is a relatively comfortable bit. Care should be taken not to have sharp edges at the butts which might cut or pinch the edges of the mouth.

Above: Single twisted wire snaffle. A degree harsher than the standard snaffle and commonly called upon when a horse charges forward or is deemed difficult to hold.

Below: The classic bar bit, felt by many to be the easiest bit on the horse's mouth. With a light line, and a quiet relaxed horse, the author has found that this bit may clank against the teeth.

Above: Double twisted wire snaffle. Notice that the joints are offset. This bit causes a pinching action in the mouth and should be considered severe.

Not pictured is the Mule bicycle chain bit which uses the chain as the full mouth piece. It is one of the most severe bits.

Below: Twisted bar bit.

Showing the proper fit of the bit in the mouth

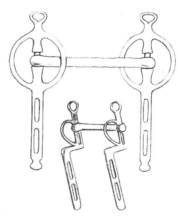

Left: There is a wide range of curbed shank bits which employ two lever principles to apply pressure. When a curb strap or chain is employed, under the chin, and the lines are hooked to the bottom most position, tremendous pressure is applied to the chin and lower jaw. Pressure is also applied to the pole or top of the horse's head. It is possible to hook this style of bit without curbs and directly at the bit ring, thereby making them quite comfortable for the horse. These bits are commonly used on show horses and fancy coaching outfits. The next two pages show many design variations.

Below: A saddle snaffle bit with oversized rings. May be useful with work horses that insist on holding their mouths wide open and prone to getting the bit ring in the mouth.

The above bits may be found with copper appointments (including copper wire twisted around the mouthpice and rollers), hard rubber wraps, and oversized (or eggshaped) butts.

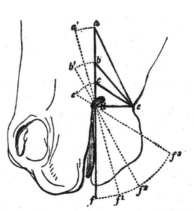

Diagram to show action of curb bit.

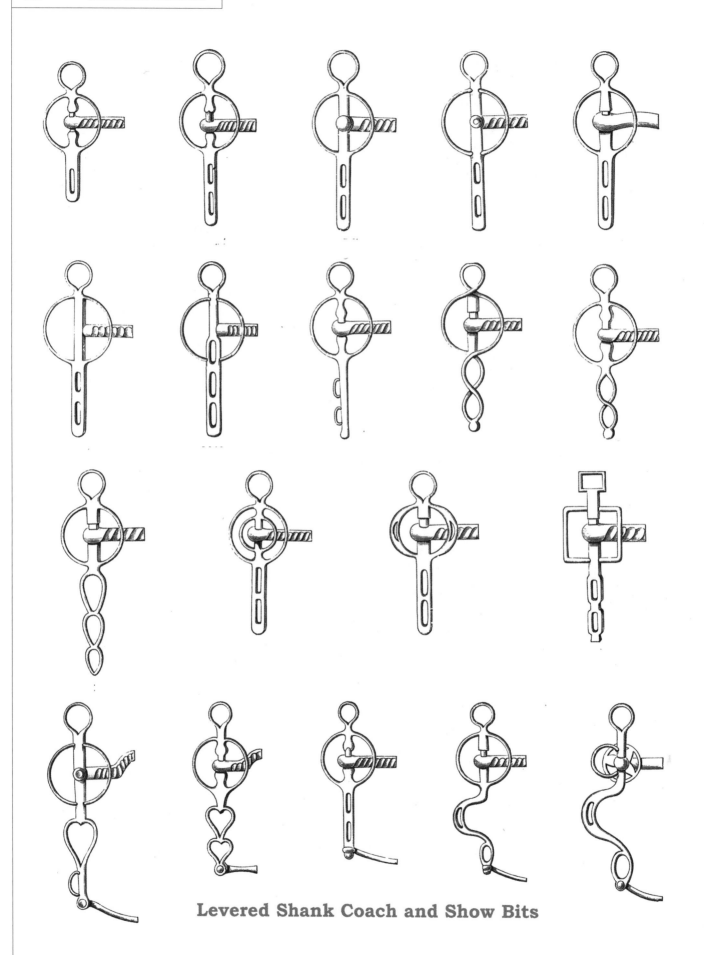

Levered Shank Coach and Show Bits

Levered Shank Coach and Show Bits

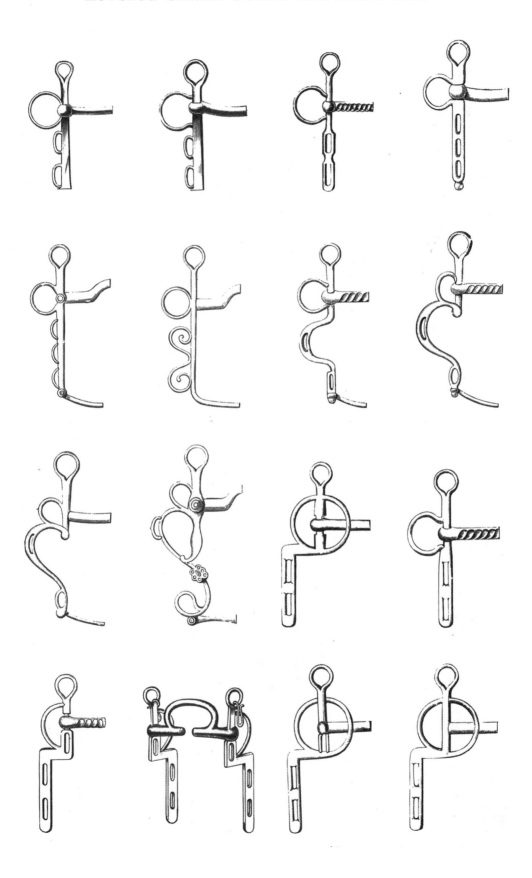

SPREADERS

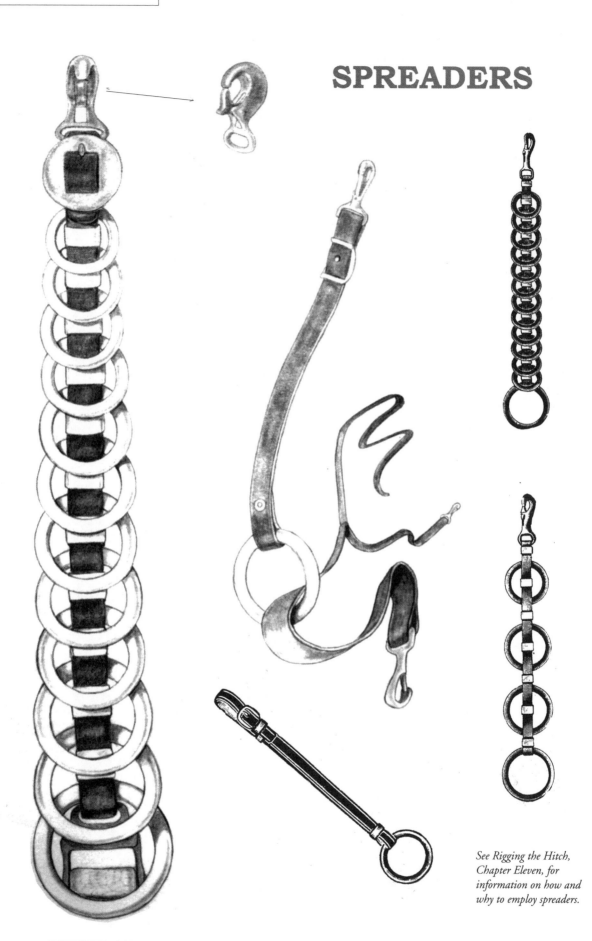

See Rigging the Hitch, Chapter Eleven, for information on how and why to employ spreaders.

ence in bits is for the common snaffle and a few variations. The illustrations included here will give some idea of the range of designs and the captions will help to explain the effects of these different bits.

spreaders

It may be desired to have a team of horses walk further apart from each other. This can be accomplished by using line spreaders (see page 141). No adjustment is made in the actual lines. The cross checks are passed through the end of the spreader rather than the inside top hame ring. The spreader itself is fastened near the top of the inside hame of each horse, there being two spreaders commonly used with one team. At times teamsters will get fancy and use four spreaders, with two per horse, but this serves no purpose other than decoration. You may have to use a longer neck yoke and double tree if you move your horses further apart, see RIGGING THE HITCH. Spreaders and spreader rings lend themselves readily to fancy design and are used frequently in western harness to dress up the horses.

center line drops

The center line drop is just a large ring, sometimes with a fancy bit of leather design fastened to it, through which both center (cross check) lines pass on their way to the horse's bits, see page 141. This is used primarily for fun but it does serve also to keep the lines organized and it is felt by some that it prevents the lines from getting tangled so easily.

shaft loops and hold back straps

These are harness parts found only on single horse harness that is to be used between shafts. For more information and illustrations see the section on single horses in the RIGGING THE HITCH chapter.

fly nets

Before the advent of fly sprays, one popular way to give the horses an aid with flying insect aggravation was to drape a netting of leather or cording over the harness. This would shake and shift helping to dislodge those pesky buggers. For those concerned about the unknown price tags of modern chemistry, the fly net is an excellent option.

Leather Harness Care

Leather harness may be cleaned with soap and water and, after it dries, it should be brushed, rubbed, wiped, or dipped in Neatsfoot oil or Neatsfoot oil compound. Leather lines are best treated with specialty products, such as *Ko-cho-Line* brand leather treatment, so that they don't become too slippery to handle.

Fly nets on a Big Hole, Montana, buck rake team. Photo by Kristi Gilman-Miller

Misc Harness Parts

Breast strap snap

Tug or trace iron or clip

Hame strap guide (top)

Breast strap hame ring

Sweat pad fastener

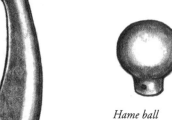
Fancy hame ball

Hame ball

Rivet

Hame housing

"D" ring

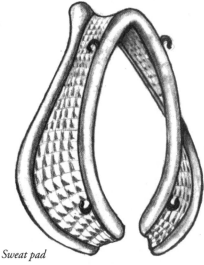
Sweat pad

Hardware

References have been made to "D" rings, snaps and conway buckles. These are just a few of the different hardware parts found in harness construction. Included on these two pages are some drawings of these parts to help you identify them.

Halter

Hame housing: *Usually made of heavy grade leather, these fit over the top of the collar with the hame tops fitting through slits. The housings protect the horses' necks during rain and snow.*

Sweat Pad: *Usually made of canvas stuffed with hair, or felt, or synthetic cushion material. The sweat pad fits under the collar. It is used for various reasons, including horse comfort, making a too-large collar fit, protecting a sore, etc. This author has come, over time, to favor the use of pads.*

Heart-shaped Center Line Drops

HARNESS HARDWARE

Special D ring

Check rein guide

Hip drop ring

Trace carrier

Square D

Hame ring with stud

Line snap

Hame stud

Hame clip rivet type

Hame clip bolt-type

Hame stud

Conway buckle

Conway broad-faced

Tug chain D

Line-end conway

*Bolt & roller assembly
Breast strap snap*

Breast strap iron

Additional Information

Hame Styles

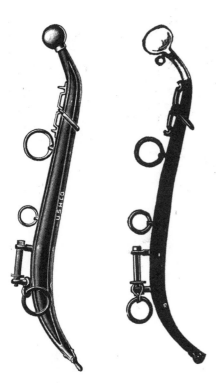

There are many styles of hames. Today the tube-steel variety represented by the three illustrated at the top of this page are the most common. On the left, notice that the clip which holds the hame strap may be slid from one notch to the next to affect a two inch range adjustment. On the right the hame straps must be threaded through any one of the fixed openings for suitable position.

Again on the right the trace or tug clips wrap around the bolts on the hames, while on the left the adjustable trace ring receives a hook or ring end tug. The style of hame on the right was prevalent in the southern U.S.

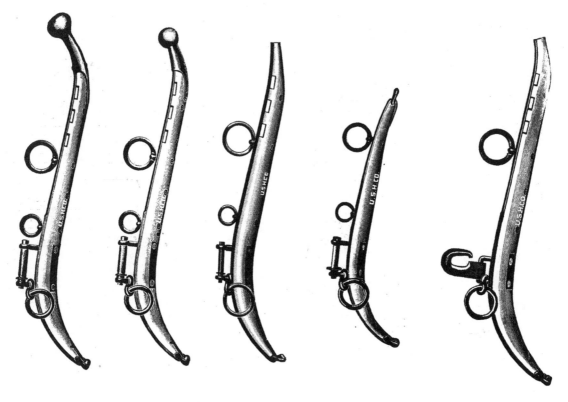

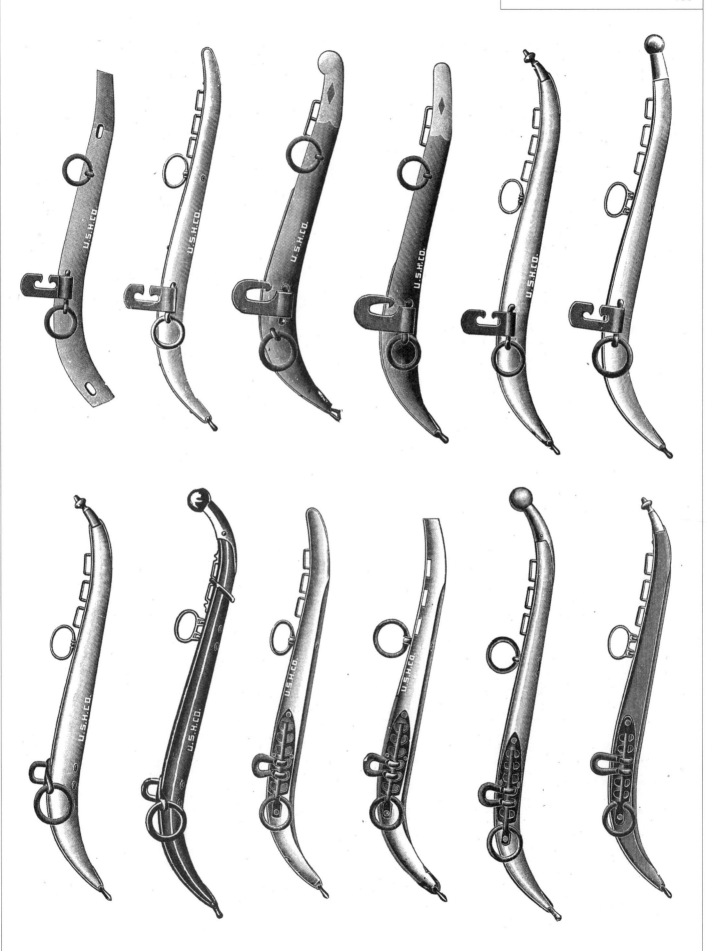

Harness Fit

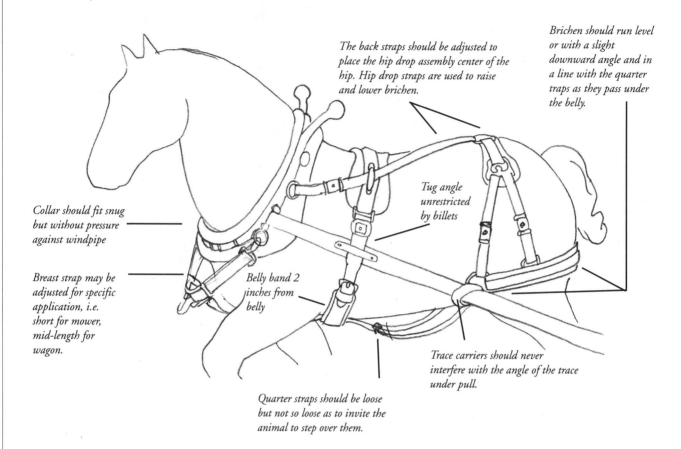

The back straps should be adjusted to place the hip drop assembly center of the hip. Hip drop straps are used to raise and lower brichen.

Brichen should run level or with a slight downward angle and in a line with the quarter traps as they pass under the belly.

Collar should fit snug but without pressure against windpipe

Tug angle unrestricted by billets

Breast strap may be adjusted for specific application, i.e. short for mower, mid-length for wagon.

Belly band 2 inches from belly

Trace carriers should never interfere with the angle of the trace under pull.

Quarter straps should be loose but not so loose as to invite the animal to step over them.

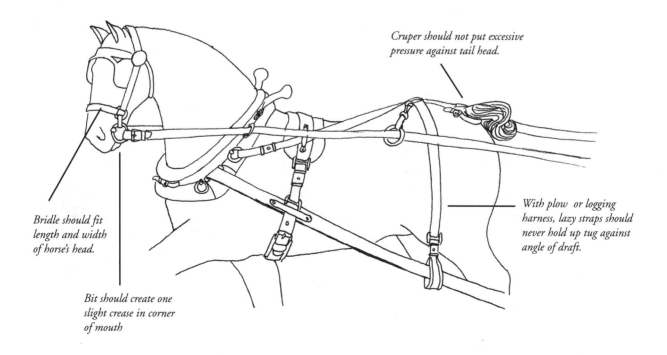

Cruper should not put excessive pressure against tail head.

Bridle should fit length and width of horse's head.

With plow or logging harness, lazy straps should never hold up tug against angle of draft.

Bit should create one slight crease in corner of mouth

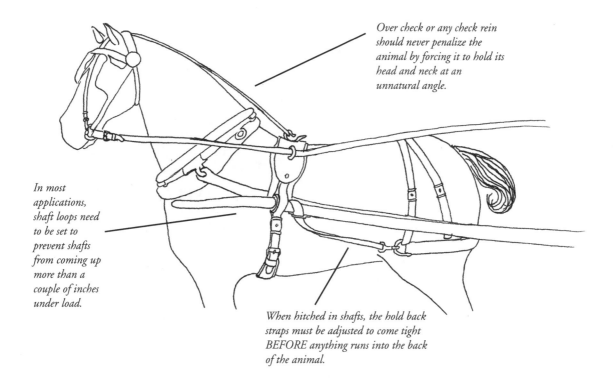

Over check or any check rein should never penalize the animal by forcing it to hold its head and neck at an unnatural angle.

In most applications, shaft loops need to be set to prevent shafts from coming up more than a couple of inches under load.

When hitched in shafts, the hold back straps must be adjusted to come tight BEFORE anything runs into the back of the animal.

Fine Tuning Team Lines

We know that standard team lines (represented in the drawing by the solid black line) work, and we think we know how they work. We also can see that the use of spreaders, effectively altering the point at which lines "cross over", allow the same two horses to walk further apart. (The spreadered lines are shown in the drawing as dashes.) But I wish to make a case for a third possible dynamic.

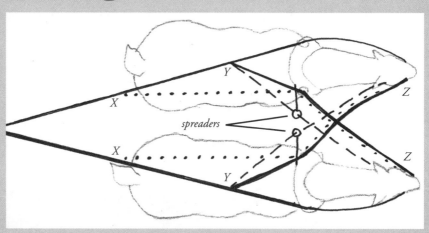

spreaders

I have not seen anything written on this particular comparative theory regarding team lines. Doubtless there will be those who will contest my conclusions.

By using longer than customary cross checks, represented by dotted lines in the drawing, we change the turning dynamic. My experiments satisfy me that half as much draw (or pull) on one line will effect the same turn when the cross checks are lengthened as shown. Studying the geometry we can

see that pulling back from point X will take less distance to effect a certain pressure at point Z than a pull back (same trajectory) at point Y. This would explain how it is that English and Canadian plow champions, with longer than usual check lines, would appear to control their teams with far greater subtlety. For many complex field hitches where effortless and absolute precision is **not** required, the shorter cross check should remain an obvious preference. They offer far less opportunity to get lines hung up.

Synthetic Harness

Today, in many Amish farming communities, a casual glance would suggest that synthetic harness (meaning harness made of nylon and/or bioplastics) has, for them, almost completely replaced leather. And again on the surface, popular notions would suggest that the reasons are obvious and simple: costs less, lighter weight, lasts longer, just as strong and it's easy to clean.

Not so simple, and in many cases most definitely not true. Synthetic harness **might** cost every bit as much as leather, can be almost as heavy, may not last as long, may not be as strong and can be downright difficult to clean. And all these things are greatly affected by just what materials are used and how the harness is constructed.

With leather harness, it is definitely true that strength and quality vary dramatically depending on the leather used, the thickness employed, the nature of the stitching, and the strengths and weaknesses inherent in design features (i.e. tapered leather strap ends punched to seat in cheap conway buckles versus sewn and riveted buckle billets). Which is to say that all leather harness is NOT the same.

The exact same observation needs to be made of synthetic harness and perhaps expanded upon. The arena of synthetic harness design and construction is still dynamic with dramatic changes sure to be just around the corner. To simplify our discussion, here are five material types being used today:

1. Nylon (webbing)
2. Polypropylene (webbing)
3. Polyester (webbing)
4. Solid polyurethane (trade name *Betathane*)
5. Bioplastic construct (trade name *BioThane*)

Nylon and polyester are the strongest material but absorb dirt and fray. (They are also abrasive to the animals.) Polypro is a cheaper, lighter material which lacks the strength of the others. Betathane and Biothane are two different and separately registered

A combination harness photographed at Horse Progress Days 2003. Biothane, nylon webbing and leather parts have been put together to maximum advantage.

trademarked synthetics. BioThane is polyurethane coating over a polypropylene webbing. With some confusion and conflicting claims within craft and industry representations, *Bioplastic* refers generally to a combination material which features a poly or vinyl coating over nylon or polypro webbing. Bioplastic has recently become the most popular synthetic for harness construction. Though the bioplastic harness is easier to clean, it apparently is not as durable as leather. The coating may crack under strain and separate from the webbing inside. Some harness makers and polymer engineers severally make claims of unexcelled washability, durability, and strength which we are unable to prove or disprove. Which is to strongly suggest that not all bioplastic is the same either as to cost, weight, durability, strength and general attractiveness.

Harness makers are finding that combinations of bioplastic and leather, employing slightly different buckling techniques and better quality hardware, allow an opportunity for top quality and strength with some construction savings.

It may seem an unfair comparison but I feel it is worth making: For a couple of decades now, manufacturers have been constructing saddles of plastics. And saddles are still being made of leather. Many people involved in performance riding would not consider plastic riding gear as they believe it would detract from comfort, flexibility and mobility. Lest it seem I am making an outright case against plastic harness, let me observe this: Outstanding

Robert Yoder of Mt. Hope, Ohio drives a four up of young Belgians in Biothane harness. Summer 2003 HP Days.

harness is being made of synthetic materials. It is not cheap. And it will benefit from proper care. Exceptional teamsters are putting this type of harness to work everywhere. It may be the sort of harness which works best for you. Leather works best for me.

With synthetic harness the reputation of the harness maker may be paramount. For the lay person to accurately assess the full quality aspects of a bioplastic harness may be difficult or impossible. It then becomes critical that the purchaser be able to trust the company or individual who has made the harness.

Mose Hershberger holds the six Beta lines for his 12 horse hitch.

beta lines

While many teamsters are holding on to leather for their harnesses, some of them are turning to synthetic driving lines called Beta or

Ben Raber holds leather team lines.

Jake Yoder's Percheron in a combination sidebacker-style harness with plastic collar pad, at Ohio Horse Progress Days.

Betathane or Beta Vinyl or Beta series BioThane because of the suitability of the material. Early bio lines were slippery, squeaky, and awkward to handle. The Beta lines (see photo this page) have a fine grained tooth or texture (sometimes referred to as a satin finish) which provides excellent grip. They appear to be strong with a minimum of elasticity. We don't have any long term experience to measure how this material will hold up under prolonged use. Where these lines may be subjected to 100 degree plus temperatures and contact with solvents, will they deteriorate? What happens when they are frozen and thawed repeatedly? Above and beyond the claims of the manufacturers, time will tell.

Harness Style Variables

The names given for these harnesses come direct from harness catalogs.

Trap or Cart Harness with French Tilbury Shaft Tugs

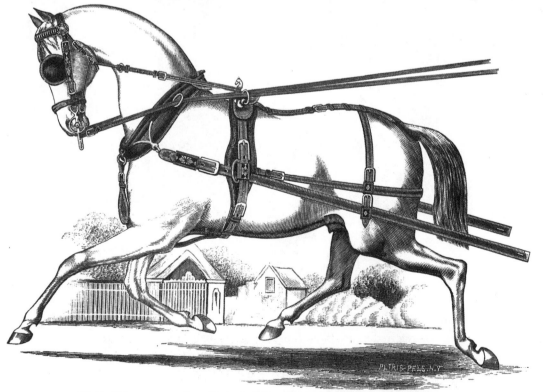

Cab Harness for Bent Shafts with Stop Plates

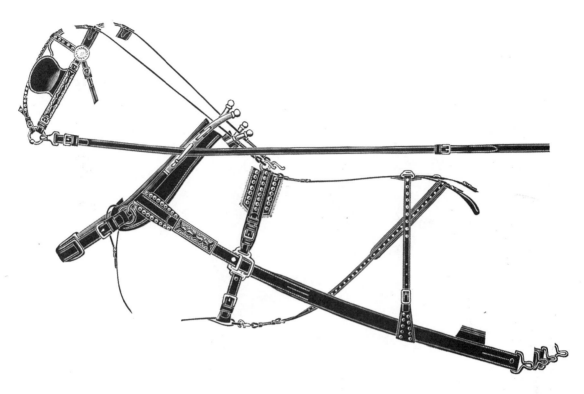

Double Spring Wagon Harness with Yankee or Hip Brichen

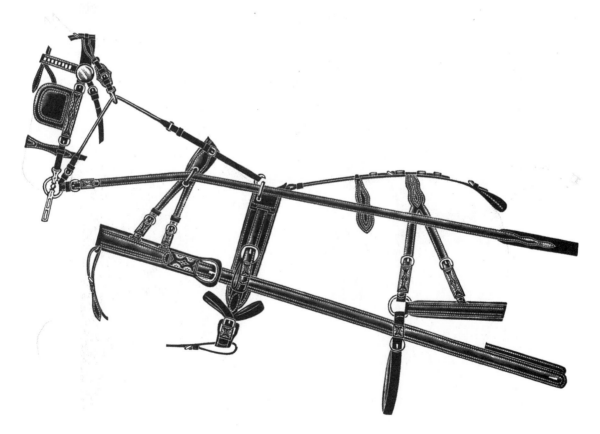

Breast Collar Swiss-style Single Surrey Harness

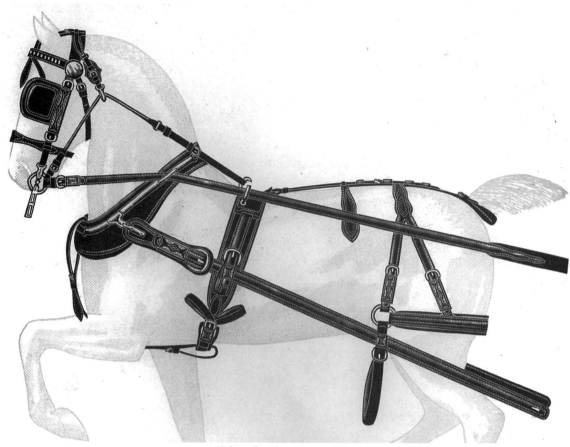

Single Surrey Harness

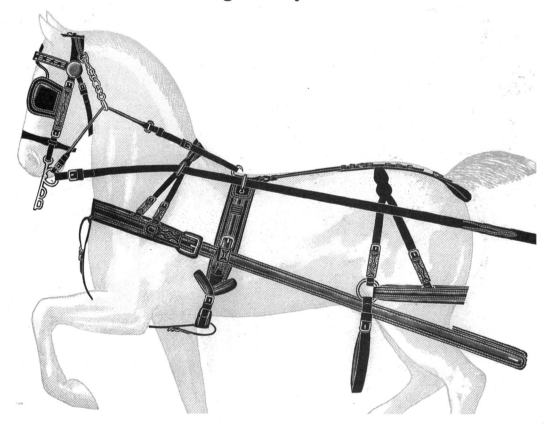

Swiss Breast Collar Single Surrey Harness

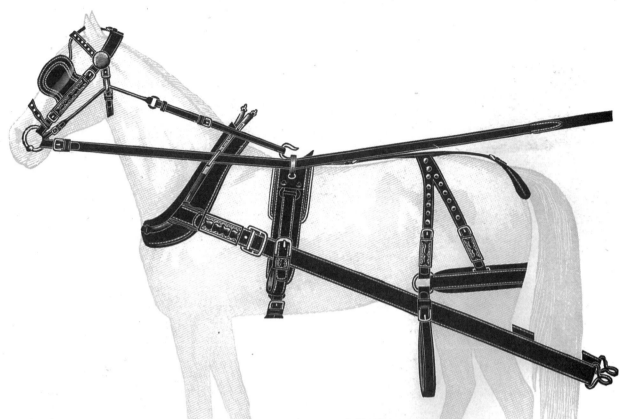

Single Express or Peddlar's Harness

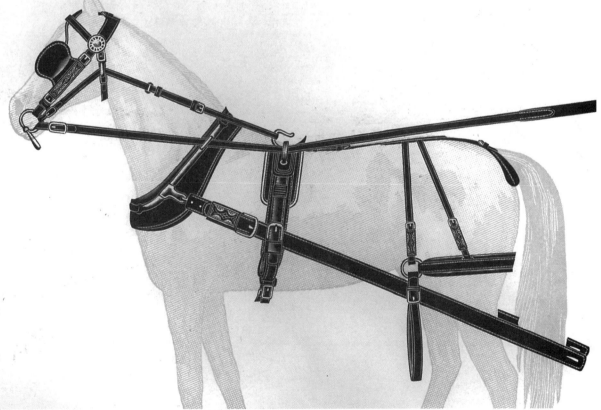

Single Grocery Harness

Two Variations of Fancy Team Coach Harness

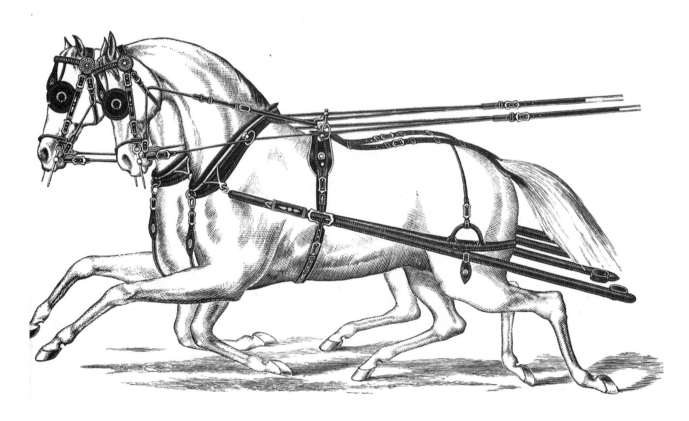

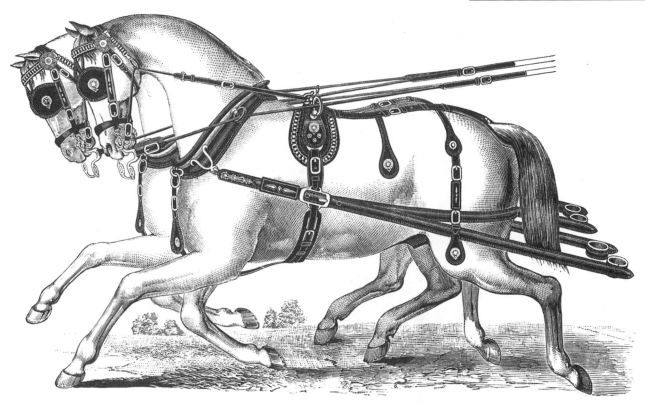

Fancy Team Coach Harness

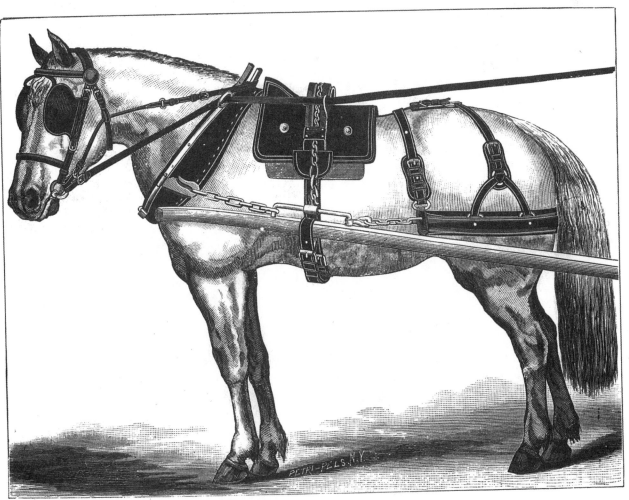

Plain Heavy Cart Harness

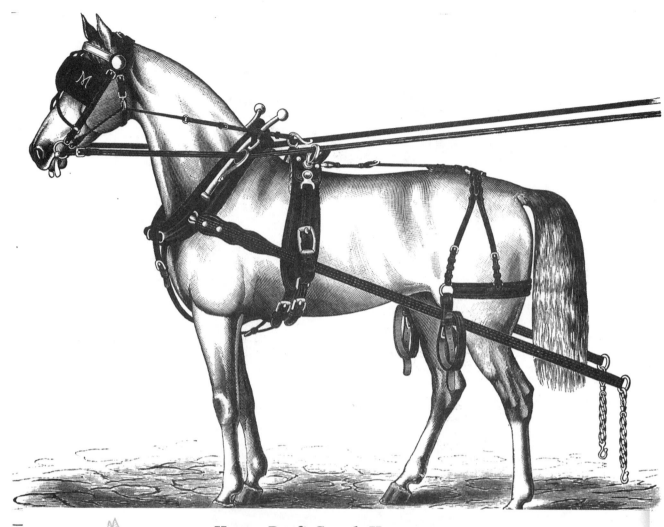

Heavy Draft Coach Harness

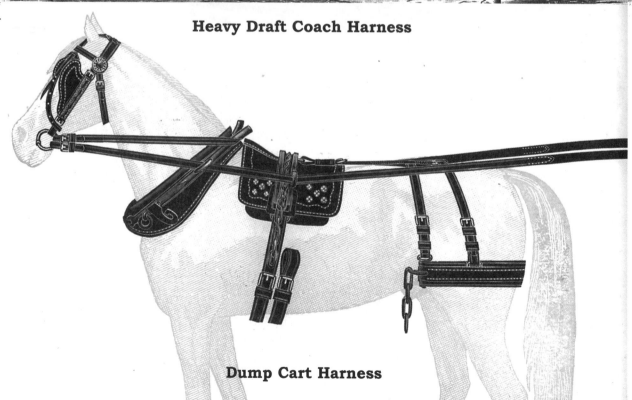

Dump Cart Harness

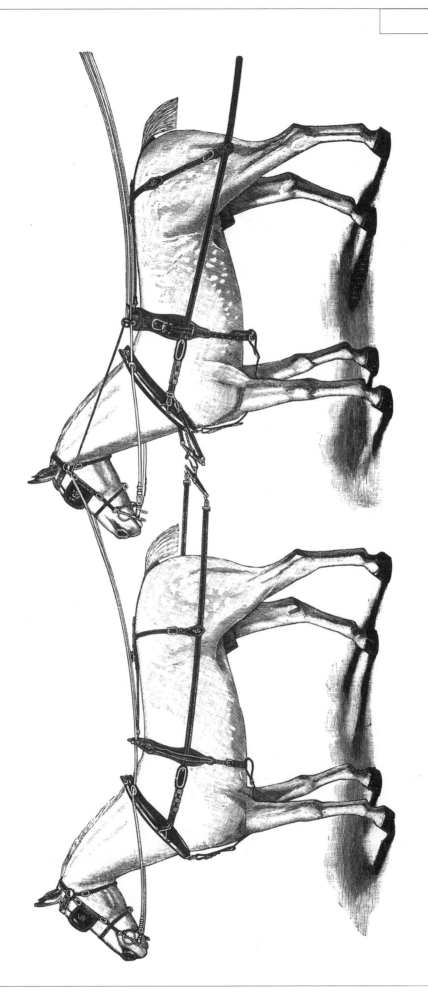

Tandem Coach Harness

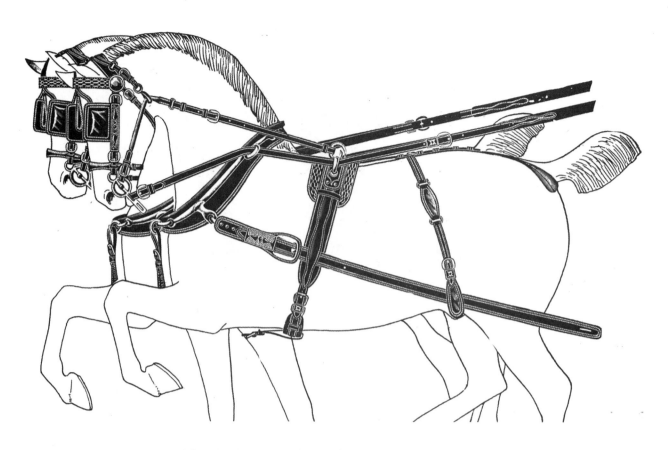

Short Tug Double Coach or Hack Harness

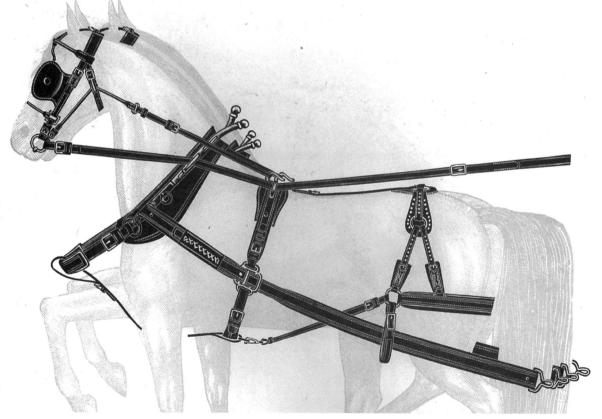

Double Spring Wagon Harness

Double Chain Southern-style Log Harness

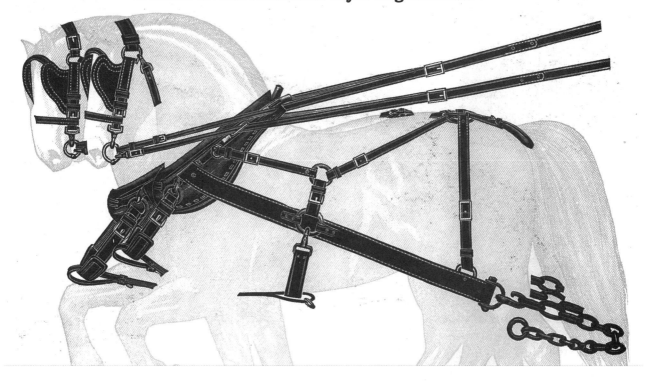

Butt Chain Market Tug Harness

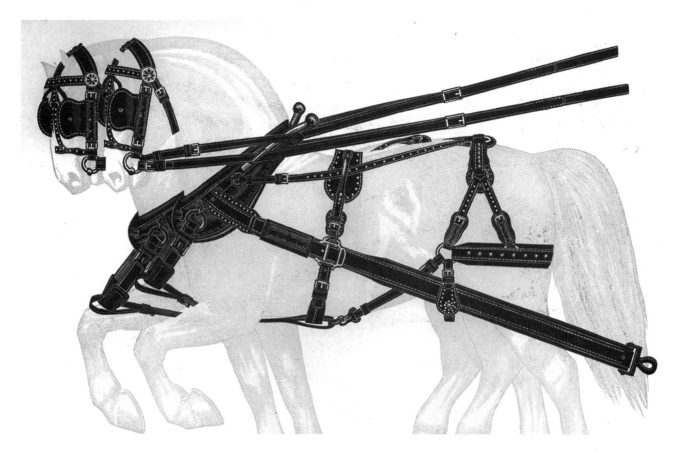

Fancy Concord Breeching Team Harness

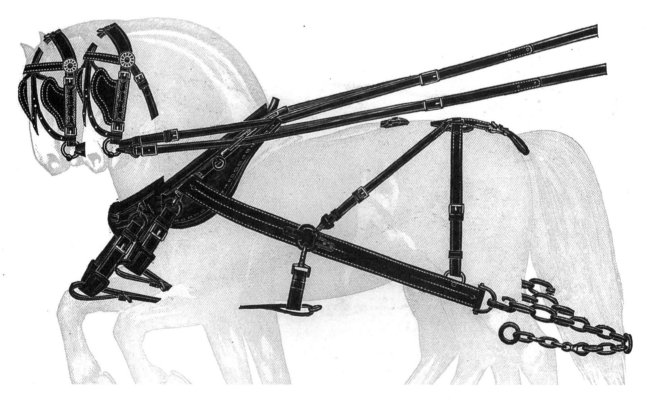

Butt Chain Harness with Double Turnbacks

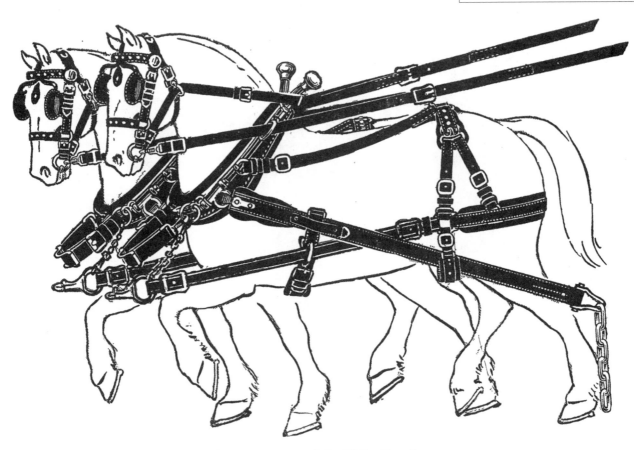

Truck Harness with Side Backers

New England or Boston Sidebacker Harness

How the New England Harness Works

Two piece tugs and two piece sidebacker straps (or jack straps) share a common side ring connection. When the horse is hitched and the forward side strap is tightened, the tug and backer straps (the black x) tighten and take all weight off the horse's back. A basket framework, see dotted lines, is formed which accounts for the possibility for incredible precision of backing, with very little sway on the tongue end.

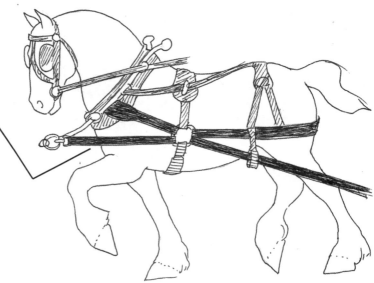

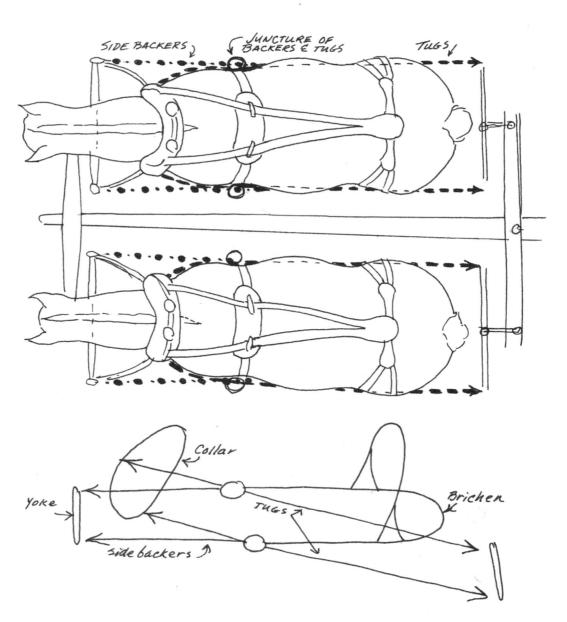

Home Harness Care & Repair

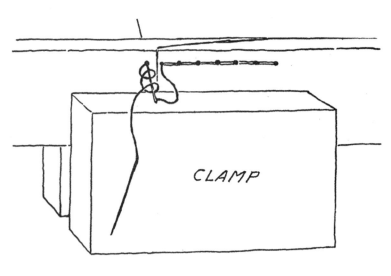

CLAMP

Simple leather harness repairs can be made at home. Waxed thread, an awl or hole punch, and suitable large needle form the rudiments of the required tools. Riveters and rivets may be purchased at most hardware stores and leather supply houses.

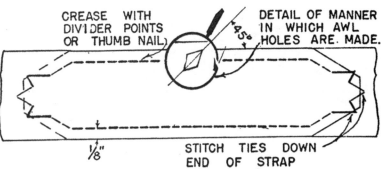

CREASE WITH DIVIDER POINTS OR THUMB NAIL)

DETAIL OF MANNER IN WHICH AWL HOLES ARE MADE.

45°

⅛"

STITCH TIES DOWN END OF STRAP

A STITCHED SPLICE

Above: An example of how to splice a leather strap with overlapping ends stitched.

¾   1¾"   1"

PUNCH TWO HOLES

CUT OUT LEATHER BETWEEN HOLES

BEVEL END

B

Above and right: How a strap is prepared to receive a buckle end. And how a buckle is attached.

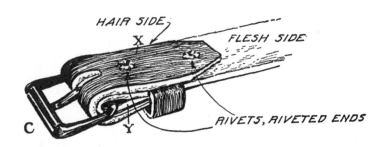

HAIR SIDE

FLESH SIDE

X

RIVETS, RIVETED ENDS

C

Y

What's a Walsh Harness?

In 1914 James M. Walsh, a government bureaucrat, designed a revolutionary harness, one with no buckles, believing that it was possible to make a longer lived and inherently stronger harness. He succeeded and later his harness business succeeded to the point that it was arguably the largest harness maker in the world in 1918. It is one of the minor mysteries of modern commercial history that this

exceptional patented harness design essentially died with the company sometime before World War II.

The Walsh Way

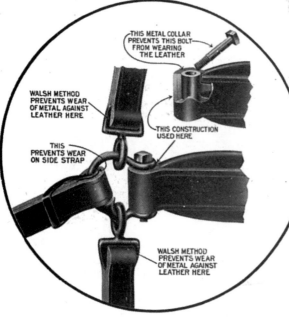

THIS METAL COLLAR
PREVENTS THIS BOLT
FROM WEARING
THE LEATHER

WALSH METHOD
PREVENTS WEAR
OF METAL AGAINST
LEATHER HERE

THIS CONSTRUCTION
USED HERE

THIS
PREVENTS WEAR
ON SIDE STRAP

WALSH METHOD
PREVENTS WEAR
OF METAL AGAINST
LEATHER HERE

The Traditional Way

Miscellaneous Notes

Three Abreast Lines

We are frequently asked for dimensional information on 3 abreast lines, the type with a second cross check. Below, in the sketch, is one suggested dimension. From where ever the line splices to six to eight inches beyond the main line, when held flat together, determines the length.

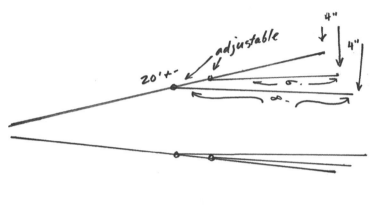

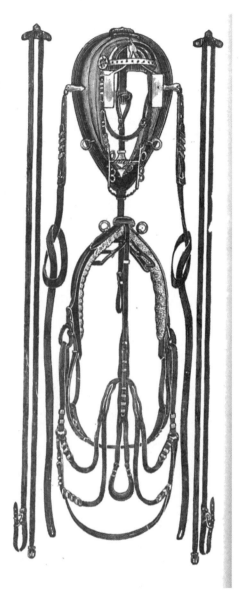

What About Show Harness?

Obviously missing from this chapter is any review or mention of Todd-style (some see or say Budweiser-style) Draft Horse Show Harness (see picture left). This book is deliberately named the Work Horse Handbook and is dedicated to the practical aspects of the horse in harness. We've intentionally pared down the content by limiting the information. There is a great deal of material available to those who are keen to learn about showing draft horses in harness. So we leave scotch tops, patent leather, tassels and swivel ornaments to others to discuss.

MEASURING YOUR HORSE TO ORDER A HARNESS

First you must identify type of horse and approximate weight. Borrowing from Smucker's Harness Shop criteria, below is a set of suggested types. Next you need to measure your animal. For some of these circumference measurements, if you don't have a cloth tape available, use a string around the horse, mark the spot, and then measure the string laying flat.

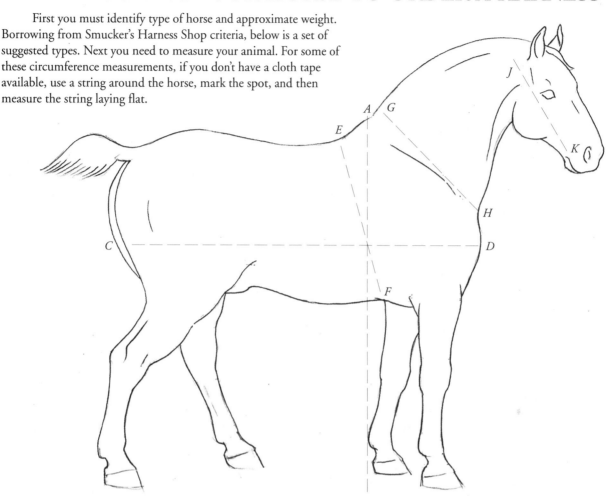

A to B is height at the withers, (or the highest point of the horsee when it has its head down to eat). For reference sake, 4 inches equals a hand. C to D is the length of the horse measured from front of shoulder to buttock on the level. E to F is the heart girth which should be measured all the way around the horse. J to K is the bridle size which measures from corner of the mouth over the poll and to the other corner of the mouth. G to H is the collar or neck size traditionally measured in a straight line from top to bottom.

NOTES: Fjords and Haflingers vary tremendously in measurements. It is not a good idea to trust to breed name for proper fit, even in the draft breeds. Measure the individuals for the harness order.

Horse Types
Small Pony 41" to 46"
Medium Pony 46" to 51"
Large pony 46" to 51"
Cob Size 700 to 975 lbs. 13 to 14.3 hands
Horse size 1000 to 1250 lbs. 15 to 16.3 hands
Oversize 1250 to 1400 lbs.
Draft Size 1450 to 1600 lbs.
Large Draft 1650 to 2000+ lbs.

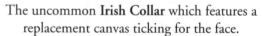

The uncommon **Irish Collar** which features a replacement canvas ticking for the face.

CHAPTER EIGHT

HARNESSING

As suggested by the previous chapter, when we speak in this book of harness we are frequently referring to all parts and pieces separately and together. This includes collars, pads, halters, lead straps, bits, bridles, as well as the main body of the harness.

The series of illustrations contained in this chapter portray one way to harness a horse. There will be individual, cultural and design differences in harnessing procedure. What you have here is a good safe way to begin. There are some practices contained here that would make good habits. It is too easy with a quiet, gentle horse to get used to bad or sloppy or unsafe harnessing habits that could cause you trouble when applied to colts or frightened animals.

Something to keep in mind: if you're new to this business, don't try to harness a horse in a restricted tie stall or confined space. If the animal should do anything unexpected, you may find yourself in danger.

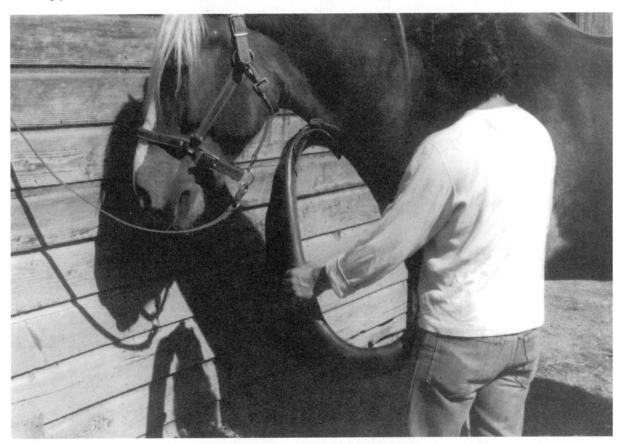

The first thing to put on the horse is the collar. If this is your first time harnessing, tie the horse to a secure spot with plenty of room for you to step out of the way. It is a good habit to take each part of the harness to the animal's head, allowing a moment for him to assess the threat. I typically put the closed collar on over the horse's head (with a few notable exceptions), this is not recommended for green teamsters or horses.

Most collars open at the top. Unfasten your collar and support both sides with your hands so that it doesn't flop open and damage the throat structure of the collar. Pass the collar gently up around the neck.

As you pass it up, do not release the outside of the collar as it could drop and crack or break at the windpipe or throat of the collar.

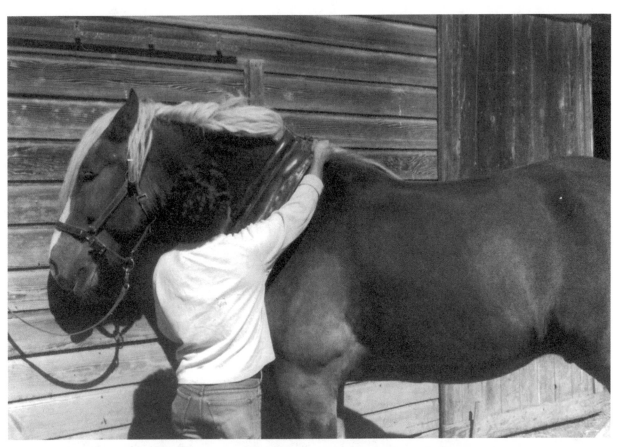

Fasten the top of the collar and pull out any mane hair from underneath.

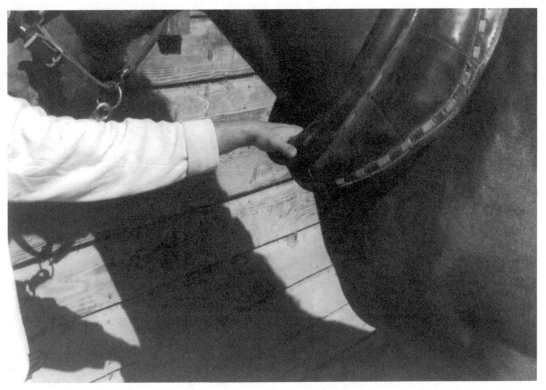

To determine whether or not a collar fits, slide the flat of your hand between the collar and the animal's windpipe. If your hand won't fit, it's too tight, if two or three hands will fit, it's too loose. Over time I have come to prefer a snug fit.

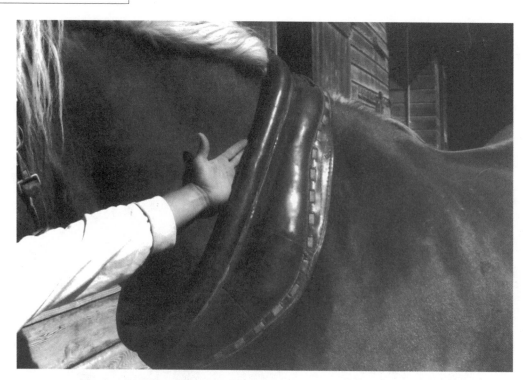

Also check the side fit of the collar. Pulled over, there should be barely enough room for a couple of fingers to slide in between the neck and the collar.

The harness in storage should hang with the hames on the left side as you face it. To pick up the harness, pass your right arm under the top center of the hip-drop assembly and on forward under the back pad. With your left hand, put the hip drop assembly on up on your right shoulder. Do the same, if possible, with the back pad. Now pass your right arm on a little further and take hold of the right hame. With your left hand, grab the left hame and lift the harness.

Carry the harness. By lifting the right hame high, as my brother Tony is doing in this picture, you should be separating the harness' left and right sides.

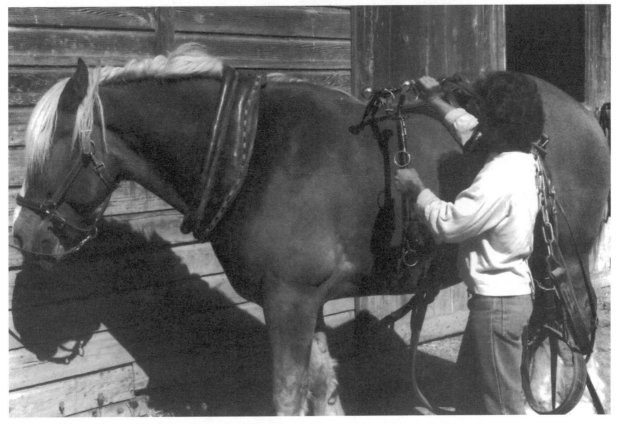

With the right hame high, and with the horse's attention, pass the right hame over the horse's middle back and move forward, towards the collar.

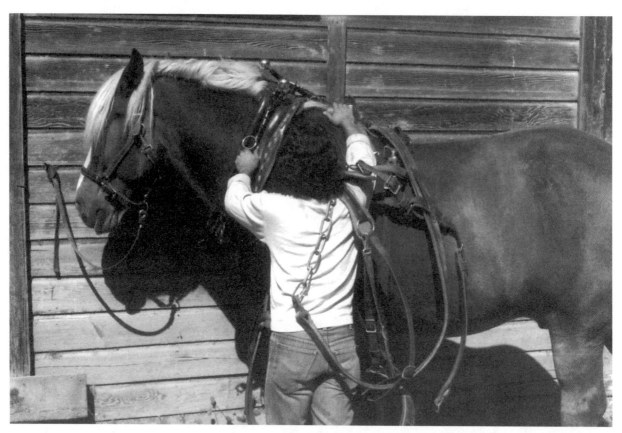

Pass the hames on over the collar while still carrying the bulk of the harness on the right shoulder.

Now, pass the back pad over the back, working the right side tug over...

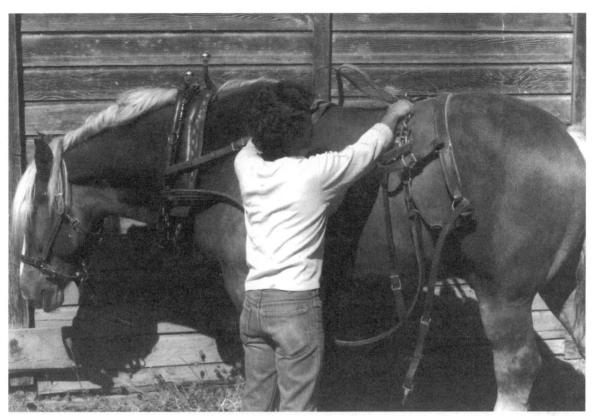

...and finally, the hip-drop assembly and brichen. Leave this rear portion of the harness up on the top of the back until positioning and securing the hames.

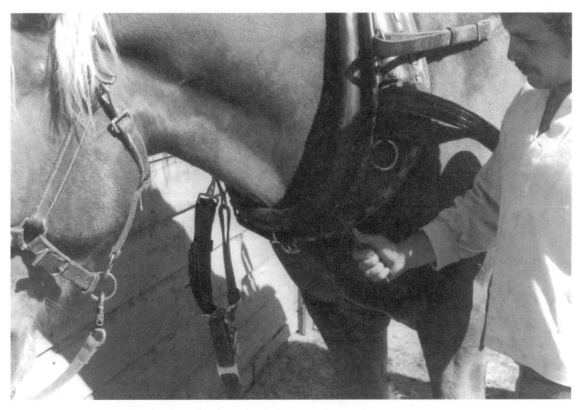

Seat the hames in the collar groove. (The top hame strap should remain fastened and the bottom strap is used in harnessing and unharnessing.) Make sure the hames are centered on the collar so that the point of draft (where the tug fastens) is correct and equal on both sides. Then fasten the hame strap...

...and make sure it is tight, very tight.

After the hames are secured, pull the harness back and in the right position with the brichen down over the tail.

Pull the tail out from under the brichen.

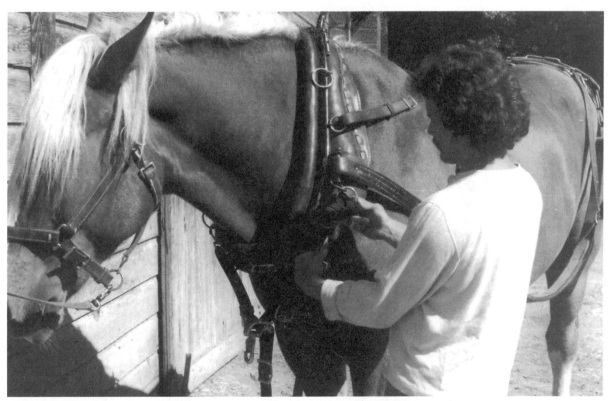

If you are using a "western brichen" team harness, snap or buckle the breast strap into the bottom hame ring of the left side.

(Notice that the collar has fallen forward just enough to give the impression of a loose fit. When the tugs tighten with a load to pull, this collar will seat back where it belongs.)

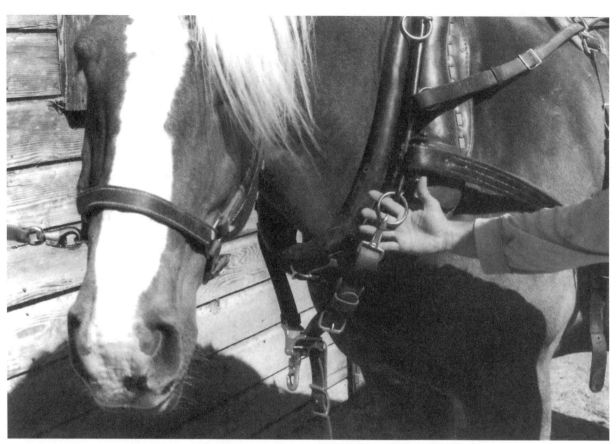

Showing the breast strap snapped. This snap faces outwards which, though serviceable, may get you into trouble should a horse rub its bit or bridle against it. It's a better idea to have the breast strap snaps face in.

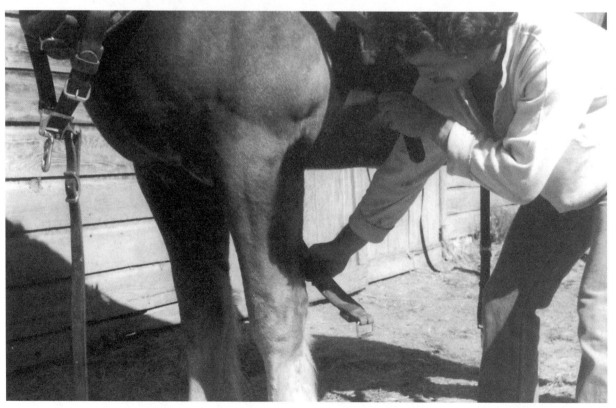

Reach under for the belly band and pull towards you...

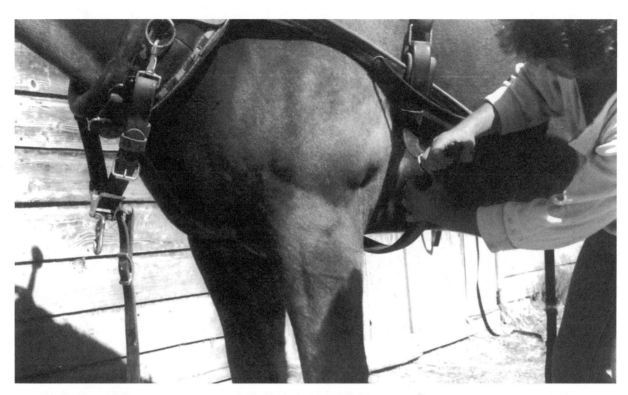

... and buckle into the belly band billet with slack. If you were to buckle it tight against the horse's belly it would serve no purpose except to discomfort the horse.

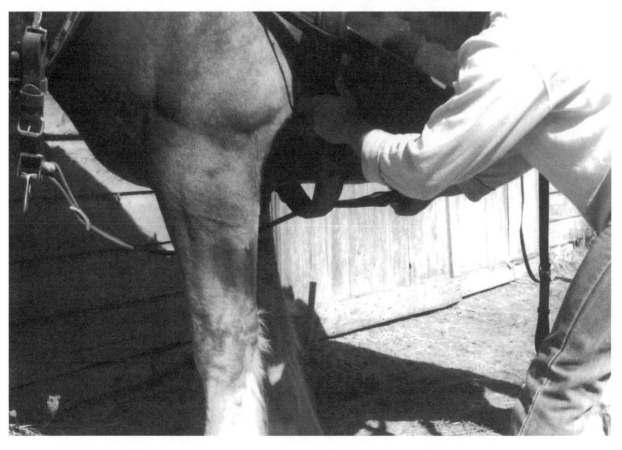

Then pull the pole strap back and over the belly band. (Note: in some harnesses the belly band passes through a loop in the end of the pole strap. See Harness Chapter.)

Snap the quarter straps (one from the front end of both sides of the brichen strap) forward into the end of the pole strap. Note that these snaps face away from the body. This reduces any chance of a sharp edged irritation to the belly of the horse.

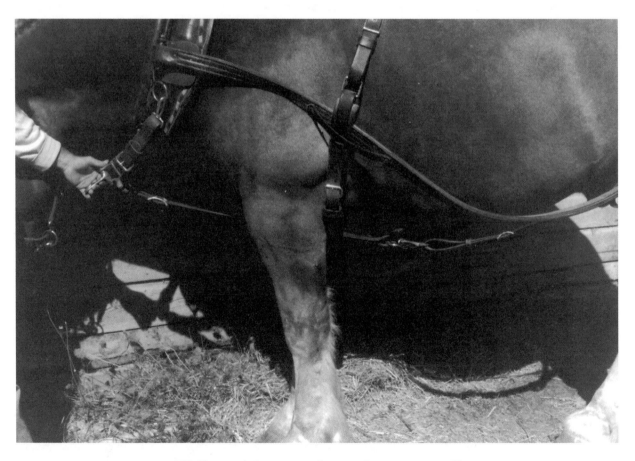

This illustrates the breast strap, pole strap and quarter strap assembly pulled tight as it would be in backing. Again through the pole and neckyoke, this also comes into play as the breaking system of the harness.

(left) Hold the top of the bridle with your right hand spreading it to fit the width of the horse's head. With the thumb and forefinger of the left hand, hold the bit wide and position it to go into the horse's mouth.

(Right) Pulling up the bridle with the right hand, put the bit into the horse's mouth. If the animal hesitates to take the bit, place thumb and forefinger in opposite corners of the mouth and gently press. The mouth should open for the bit. (NOTE: Be careful not to put a frozen or extremely cold bit into the horse's mouth.)

Lines

If you are using a team, after your horses are harnessed you need to thread and fasten the lines. Follow this diagram remembering that the cross check (or spliced line portion) must cross the center of the team after first passing through the top hame ring. The continuous portion of almost all line designs runs along the outside of each horse, through the top hame ring and fastens to the bit. Whether you are using snaps or buckle billets, make sure they face out away from the horse.

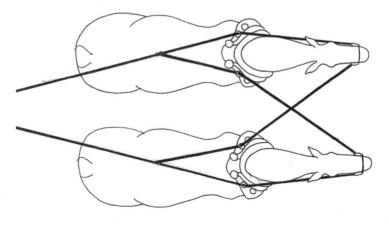

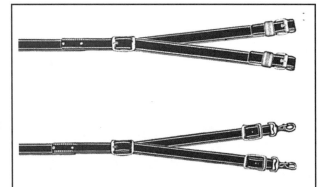

Line Snaps vs. Buckles

The front end or ends of each driving *line* (often mistakenly referred to as *rein*) must fasten to the end of a bit. The most common fastener is a metal snap. But I cannot recommend this hardware, it is not a secure way to fasten lines. Snaps break, they unsnap, and they hang up on other hardware, ropes, and misc. I prefer, and strongly recommend, buckle billets on line ends.

HARNESSING CHECKLIST

1. Always speak to the horse before approaching from the rear.

2. Check horse's shoulder for lumps or galls. Check inside collar surfaces. Make sure surface is clean. Put on collar, being careful to support both sides. Make sure collar fits properly.

3. Put on harness. Attach hame strap first. Make sure hame strap is tight. Buckle bellyband. Then pull harness brichen back into place. If brichen harness, attach pole strap, quarter straps and breast strap.

4. Put on bridle. Make sure bit fits properly and both ears are through.

5. Attach driving lines with buckles or snaps facing out.

6. Double-check all buckles, straps and lines.

Checking Harness Adjustments

Unless this horse is now harnessed with the same outfit it has been working in, now is the time to check the fit.

A. Push back the collar and double check its fit. Collar fit is critical.

B. Eyeball the position of the hames in the collar. Are they centered and well seated? Too short? Too long? Proper hame fit is not critical but it will affect efficiency and comfort.

C. Check the tightness of the hame straps to see if they are seated in the collar groove top and bottom. Loose straps could pop loose and cause problems.

D. Are the quarter straps hanging too low? If they are, it may cause the neckyoke to rise up too much when backing and allow horse to get a leg over and break a strap.

E. Is the brichen in its proper place? Not critical but definitely important.

F. Does the bridle fit the head? How do the blinders line up with the eyes? Does the bit fit? A sloppy fitting bit and bridle may come out and off. A tight fit may cause discomfort.

For guidance in proper adjustment of harness see pages 131, 140, & 141.

Old Harness Notes

With used harness, all parts MUST be strong. Be sure to check leather hidden inside of conways and buckles. These are expected results if certain parts should break:

When one bit strap breaks – bit falls from mouth, loss of control, unable to stop or turn horse.

When any part of line breaks – same results as above. May be able to turn horse(s) in one direction.

When a hame strap breaks – harness is pulled back off horse, likely to result in wreck.

If pole strap, quarter strap or breast strap breaks – unable to back up vehicle, unable to prevent vehicle from rolling up on horse's heels.

If bellyband breaks – may cause horse to balk on heavy pull.

If tug breaks during pull – horse falls forward and down, might cause injury to horse.

In particular, all of the above parts should be carefully checked when considering the purchase of used harness.

harness trouble shooting

• *Reasons why harness fit might cause a balky horse*: Pain from repeated sharp edged rubbings, broken rivets or nailings jabbing, too small collar choking off the wind, check rein too tight, choking from a lift with tugs caused by too loose a belly band, etc.

• *Reasons why harness fit might cause a nervous horse:* Any of the above will probably worry a horse and make for heightened anxious awareness. With an animal prone to nervousness this may lead to dangerous behavior.

• *Reasons why harness fit might cause a horse to lean one way or the other*: To avoid an irritation or a pain a horse will naturally lean away or move away from the offense. Classic examples include but aren't limited to; sore shoulder at collar contact, sharp edge at blinder, cutting irritation at one side of the bit, sharp edge at tug end rubbing rear leg, etc.

• *Reasons why harness fit might result in breakage:* A classic example is the too loose quarter strap, the horse kicks at flies and gets a leg over and the strap or snap must break. Improper hole punching in straps can cause ripping and/or breakage. A twisted old line, under hard pull, may rip. Too short an adjustment of lazy strap causing downward pressure from taught tug thereby breaking strap.

• *Reason why harness fit would cause horses to dislike backing:* Too long a hookup with quarter and pole straps may cause neckyoke to raise very high when backing and breaking thereby causing discomfort for the team.

• *Reasons why harness fit would negatively affect head position:* Check rein set too tight or hung up in guides and holding head sideways up

• *Reasons why harness fit would choke horses:* Collar too short. Belly band way too loose or gone causing, in some pulls (i.e. downhill irregular terrain), to pull collar up and against windpipe.

Ed Joseph harnessing. Photo by Kristi Gilman-Miller

Clayton McLaen, of North Dakota, with his team and #9 mower. Notice the homemade burlap and cord fly net.

To Unharness: Remove the bridle and, if need be, put a halter on the horse, tying him up. Unsnap the quarter straps, unbuckle the belly band, unsnap the breast strap left side, and unbuckle the bottom hame strap. The harness should be sitting free on the horse. Lift the brichen over the tail and up on the hip. Pass your right arm under the harness body passing its bulk up on your shoulder. Then reach over, with right hand, to take hold of the right hame. Grab the left hame with your left hand and back up sliding the harness off towards you. Be careful not to let the harness fall off, backwards, to the horse's hocks or heels as it could spook, especially a green horse.

If you're in a tight stall be slow, calm and cautious in taking off the harness.

The collar is the last thing to come off. Until you have the confidence to remove over the head, unbuckle the collar at the top and, once again, take care to support both sides as you let it down. It is a good pratice to buckle the collar back together as soon as it is taken off the horse. Also, when storing always hang collars upside down. The best time to clean a collar is immediately after being taken from the horse; wipe with a rag and neatsfoot oil.

Note: if you are using a collar pad you will need to unclip it from the collar on the left side before you undo and remove the collar.

A Sunday scene at a churchyard in Ontario, Canada circa 1980. Photo by Kristi Gilman-Miller.

CHAPTER NINE
HITCH GEAR

In this photo, taken at a plowing demonstration on the Thomas Ranch in Waitsburg, Washington in 1980, you see (top left) eight mules, (top right) six horses, and (center) four horses. Photo by Lynn Miller

You've got some willing horses or mules or ponies or donkeys in harness and you want to go to work. The next things you need are the *Hitch Gear*, those inbetween things, the apparatus which hooks your animals to that which you pull.

The beginning of this chapter features illustrations of traditional hitch equipment with basic descriptions of function. If you are new to the business much of what is pictured here will be confusing, at least until you see how it all might fit together to get a job done.

Unique dynamics and structural peculiarities abound when we get into the specialized application of true horse power. The particular requirements of a mower evener, the odd equalizing setup for a buckrake, the precision required of certain plow eveners, the comfort a brichen evener might bring to the job of using a single horse on a hayrope haulback, the nature of hooking to a riding cultivator, a butt chain stretcher bar for logging purposes, these only begin to suggest the variety out there. Most of it is not covered here, or if it is it's a quick passing reference. The reason is simple, too much information to pack into one book.

(This is why I have authored several other companion volumes to this one, each targeting a procedure or set of procedures and going into depth on the specifics involved including hitch gear. These books include, at this time, Training Workhorses /Training Teamsters, Horsedrawn Plows and plowing, Horsedrawn Tillage Tools, and Haying with Horses. With a little luck there will be several more books to add to what we refer to as the Work Horse Library. At this point there are 2,224 pages of information published. We expect to add another 1,200 pages.)

In the last two decades, several innovations have appeared in work horse hitch equipment. Perhaps the most significant of these has been the reintroduction of rope and pulley equalizing systems for multiple hitches. Those are covered in this chapter and Chapters eleven and twelve.

There are cultural differences and individual inventions which aren't shown here. However, these illustrations, coupled with the information in the next (and previous) chapters, should give you a sound understanding of the principles (and possibilities).

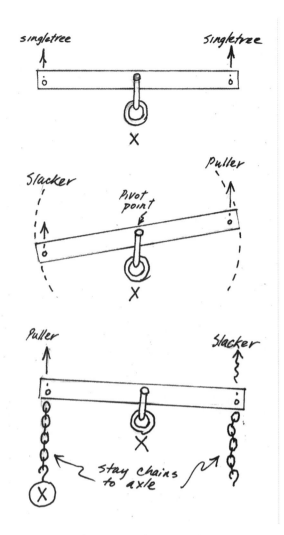

Naming the Sticks and Apparatii

We speak of *eveners* (sometimes equalizers and/or stretchers) which literally refers to the function and dynamic of requiring from our animals that they each pull their share of the load. The vernacular can be confusing. We speak of a *doubletree* as both the complete unit with evener and *singletrees* (or swingletrees) - or as just the evener stick itself. Equally when we speak of a *triple tree* we may be referring to the three horse evener bar from which everything else hangs or we may be talking about the entire unit.

We speak of *neckyokes* and the uninitiated sometimes think we are logically speaking of that leather donut we put around the horse's neck. Not so. A neckyoke is the stick which hangs in front of the team or span of animals and carries the pole or tongue. Through the pole straps, breast straps and pole, the neckyoke functions as one rigid rib of the backing and breaking system.

We speak of *stay chains* and what we mean are those chains we might use to artificially keep an evener straight.

We call the hardware pieces we use to fasten eveners together *shackles* or *clevises* depending on where you come from.

Shafts are twin rigid pole-like assemblies connected to vehicle or implement and which go on both sides of the single horse.

I have no doubt that in far flung corners there are other names to call these pieces.

Keeping the animals pulling equal

A simple doubletree (as evener stick) is built to pivot at the center where the shackle pins in. Each horse in fastened via singletree to the opposing ends of this doubletree. If both horses make a comparable effort to pull, the stick remains relatively even. If, however, one horse pulls and the other lays backs or slacks off, (see drawing on the left) the doubletree will show this in its angled position. At a certain point the puller actually is pulling the slacker back via the evener. This can cause problems. It is normally resolved by having two animals who are well matched to work together, this speaks to speed as well as strength. It often is modified by the skill and sensitivity of the teamster. There is a device we call a *buckback strap* which will aid in correcting this problem. (See page 189). And, on occasion, with wheeled vehicles where the pull is slight, stay chains may be employed. These fasten on the ends of the doubletree and pass back to hook on the wagon axle. If an auto steer vehicle is being hitched, (one on which the axle does not turn when turning) the stay chains hook back to the pole (see below). With the stay chains, there is no equalizing effect as the pulling horse works against two points to pull the entire load.

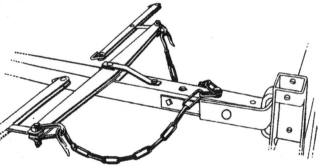

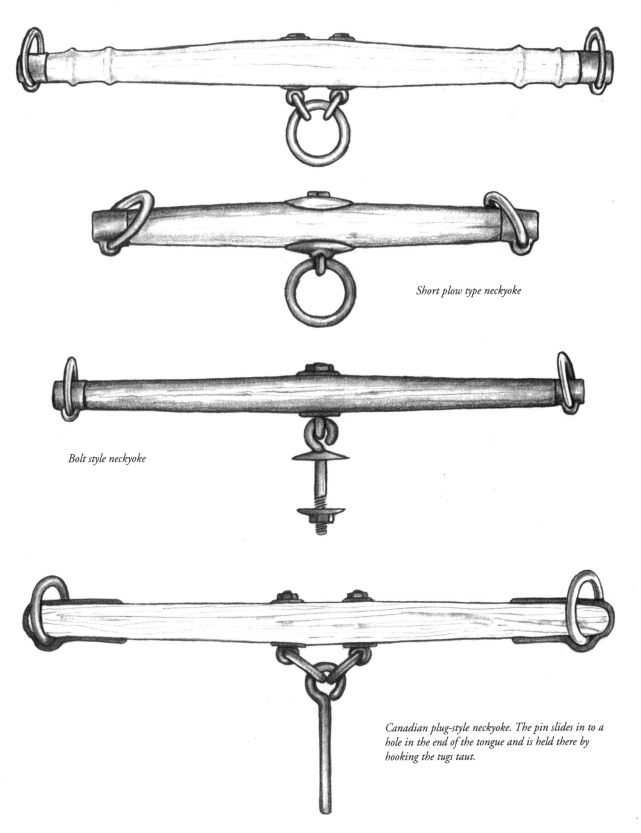

Standard wagon neckyoke - commonly available in two inch increments, from 34" to 48", to match doubletree length

Short plow type neckyoke

Bolt style neckyoke

Canadian plug-style neckyoke. The pin slides in to a hole in the end of the tongue and is held there by hooking the tugs taut.

Standard singletree dimensions are 26 through 38 inch with the longer units being used primary on wagons.

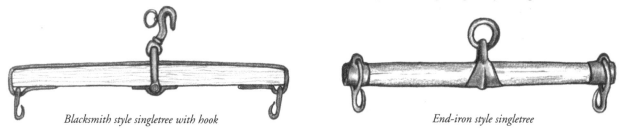

Blacksmith style singletree with hook

End-iron style singletree

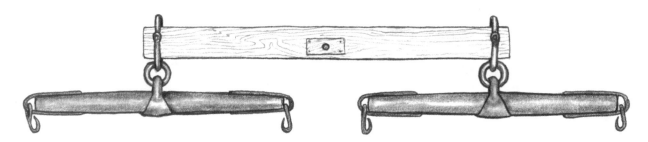

Wagon doubletree set with stay chain clevises. Standard wagon doubletree lengths for full-sized horses are 42 through 48 inch. Narrow plow-type doubletrees are 36 to 40 inches in length and mathed to width of furrow cut. See **Horsedrawn Plows and plowing** *by L.R. Miller for details.*

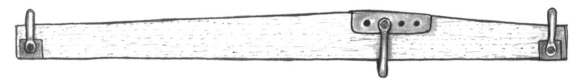

Three horse evener is 50 to 54 inches in length depending on the length of doubletree used. Hitch point is 1/3 over.

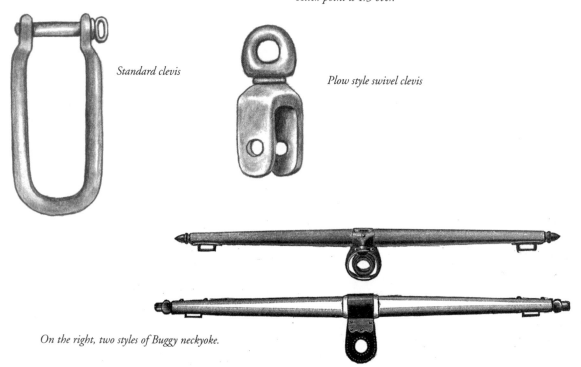

Standard clevis

Plow style swivel clevis

On the right, two styles of Buggy neckyoke.

A few of the wide variety of clevis or shackle made for eveners and hitching to implements

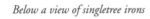

Two styles of stay chain shackles

Below a view of singletree irons

The late, legendary, Don McInnis of Oregon, drives three of his Belgians on a foot lift sulky plow with a steel centerfire three abreast evener. Don plowed well into his nineties. Photo by Nancy Roberts.

*The author, back in 1977, adjusts the length of hitch on the Oliver sulky plow. The late Bob Nygren drives Lucky, Jewell and Dick on the author's Junction City farm. Side to side, and up and down; the exact position of the evener on a plow will affect the width of the cut and whether the plow stays in the ground. For particulars on this subject, see **Horsedrawn Plows and plowing** by this same author. Photo by Nancy Roberts.*

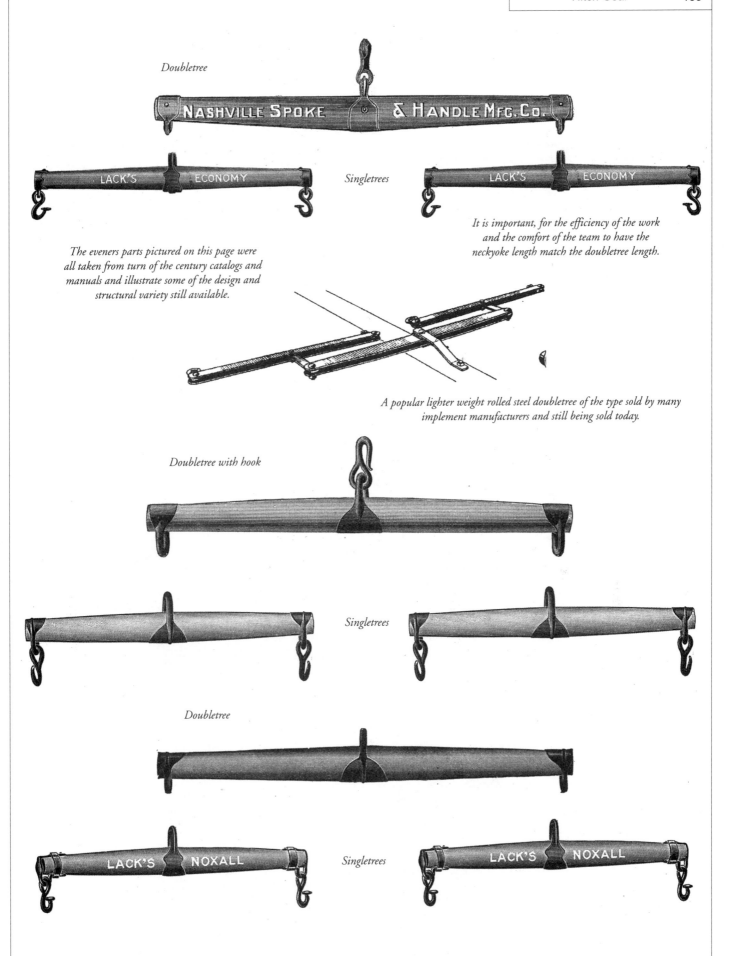

Doubletree

Singletrees

It is important, for the efficiency of the work and the comfort of the team to have the neckyoke length match the doubletree length.

The eveners parts pictured on this page were all taken from turn of the century catalogs and manuals and illustrate some of the design and structural variety still available.

A popular lighter weight rolled steel doubletree of the type sold by many implement manufacturers and still being sold today.

Doubletree with hook

Singletrees

Doubletree

Singletrees

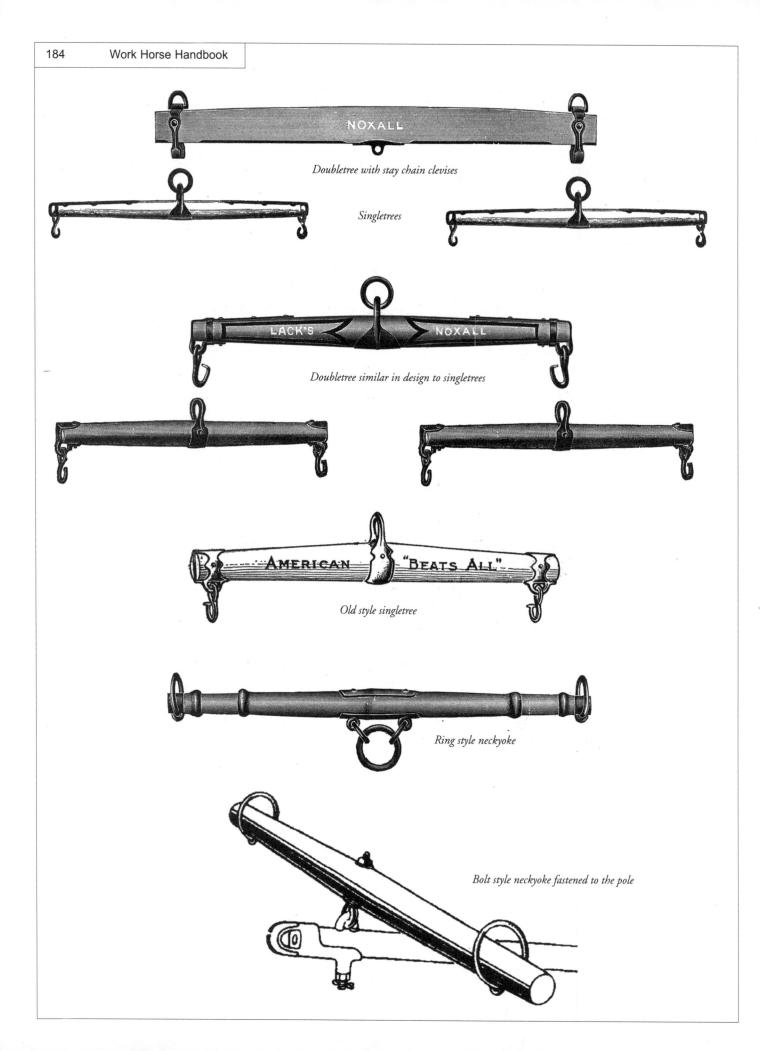

Doubletree with stay chain clevises

Singletrees

Doubletree similar in design to singletrees

Old style singletree

Ring style neckyoke

Bolt style neckyoke fastened to the pole

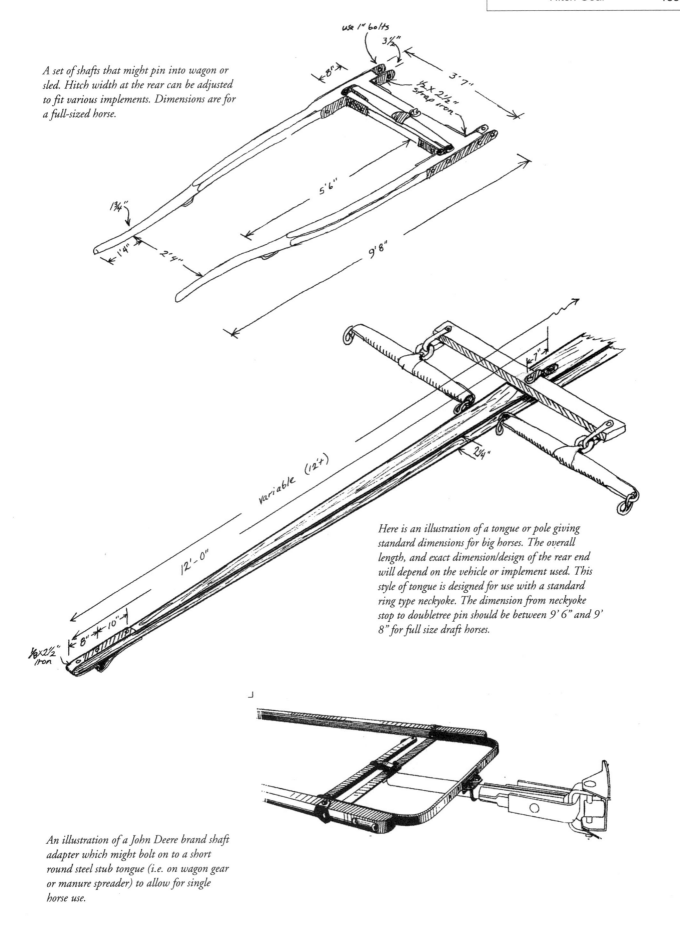

A set of shafts that might pin into wagon or sled. Hitch width at the rear can be adjusted to fit various implements. Dimensions are for a full-sized horse.

use 1" bolts

3½"

8"

3'-7"

½ x 2½" strap iron

5'6"

1¾"

1'4"

2'4"

9'8"

variable (12'+)

7"

2¼"

12'-0"

8" 10"

⅜ x 2½" iron

Here is an illustration of a tongue or pole giving standard dimensions for big horses. The overall length, and exact dimension/design of the rear end will depend on the vehicle or implement used. This style of tongue is designed for use with a standard ring type neckyoke. The dimension from neckyoke stop to doubletree pin should be between 9' 6" and 9' 8" for full size draft horses.

An illustration of a John Deere brand shaft adapter which might bolt on to a short round steel stub tongue (i.e. on wagon gear or manure spreader) to allow for single horse use.

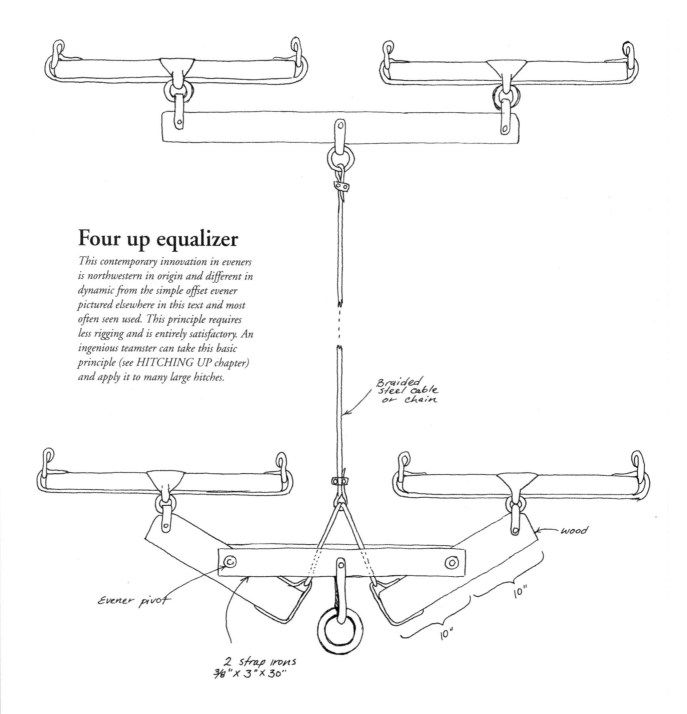

Four up equalizer

This contemporary innovation in eveners is northwestern in origin and different in dynamic from the simple offset evener pictured elsewhere in this text and most often seen used. This principle requires less rigging and is entirely satisfactory. An ingenious teamster can take this basic principle (see HITCHING UP chapter) and apply it to many large hitches.

Braided
steel cable
or chain

wood

Evener pivot

10"

10"

2 strap irons
3/8" x 3" x 30"

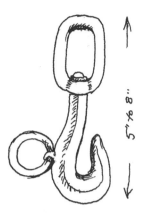

(above) The chain assembly for the stretcher bar pictured below.

(left) A yarding chain for hooking logs.

(right) A sample of one style of butt chain used on the harness of the same name.

A swivel grab hook used on logging singletrees and doubletrees

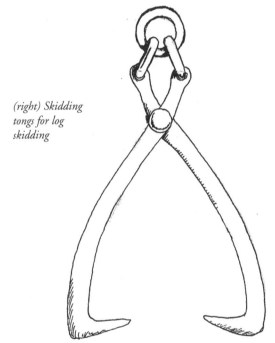

(right) Skidding tongs for log skidding

Merrill Bradshaw and his Percheron team logging in the Bitteroot forest of Montana. Photo by Tracy Mumma

(left) A 'stretcher bar' logging evener. The chain around the doubletree (bar) serves to check see-saw action and make the animals work together better. The singletrees are butt-chain type of the length preferred in the north woods. Other areas use narrower singletrees and doubletrees (40" to 30"). The illustration shows two styles of butt-chain; one can be engaged at any length for flexibility - the other hooks either full length or half length. Both hook into hooks in the end of the tugs.

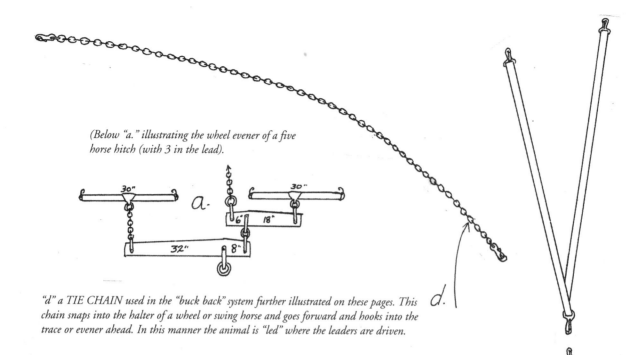

(Below "a." illustrating the wheel evener of a five horse hitch (with 3 in the lead).

"d" a TIE CHAIN used in the "buck back" system further illustrated on these pages. This chain snaps into the halter of a wheel or swing horse and goes forward and hooks into the trace or evener ahead. In this manner the animal is "led" where the leaders are driven.

(Below "b.") shows the wheel evener of a 3 x 3 six-up.

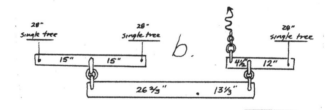

These evener illustrations are given to explain the "offset" principle which works like a common see-saw, the center point of which is moved to allow a light person to equal the weight of a heavier person. These eveners equalize the load so that all animals must pull equal.

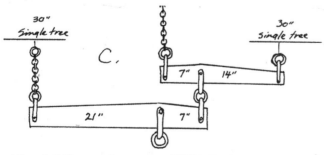

(Above "c.") illustrates the wheel hitch of a basic four-up plow hitch.

"e." A BUCK BACK STRAP. This item is best illustrated in the next chapter in the multiple hitch illustrations. It snaps into both sides of the bit on the same animal and then runs back to snap into the adjoining animal's trace or to the lead chain. In this manner, the "bucked back" horse is prevented from moving ahead unless the hitch does.

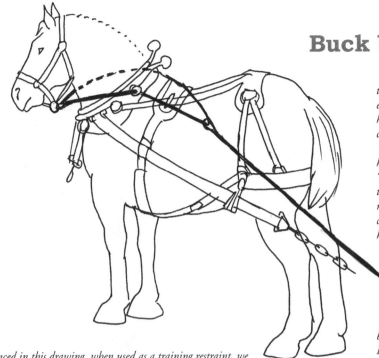

Buck back straps

It was hard to decide where to put this device, under harness? under hitch gear? elsewhere? So we've tossed the coin and put it here and also in the RIGGING THE HITCH chapter.

Originally designed as a multiple hitch device which allows that wheel horses be "Bucked Back" and "tied in" so that the teamster only has to drive the leaders, we recognized its dynamic value as a training or correctional restraint for green and/or anxious horses as well.

As evidenced in this drawing, when used as a training restraint, we typically snap both ends into the halter, worn under the bridle, and the other end to the teammate's singletree. When the buck strapped animal attempts to charge forward or take the entire load, pressure is put on the nose and he must back off. When properly adjusted, this strap remains slack, if both animals are pulling even.

As shown also on page 188, it can be made of leather or rope. (My good friend Kenny Russell of Mississippi makes his of a Beta or synthetic material.)

When used as part of the multiple hitch, see chapter eleven, and snapped into the bit rings the Buck Back Strap becomes a 3rd dimensional connection for driving big units.

Over time we have found more and more opportunities to use this simple rigging. And we recommend every serious teamster keep one or two around. It is a most humane way to solve what can be a vexing problem in but a few horses.

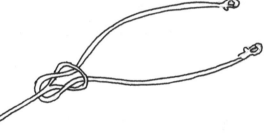

A buck back made of rope braided into snaps

Hooking Butt Chains

In one of his **Ask a Teamster** columns published in **Small Farmer's Journal**, Dr. Doug Hammill went into great depth discussing the hows and whys of butt chain traces. His drawing below demonstrates how the ring style chain is doubled back to the trace, halving the distance the horses are from their load. A handy step when pulling logs and wanting to add lift.

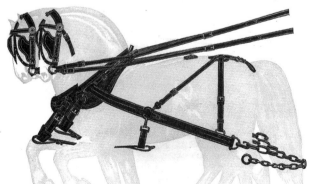

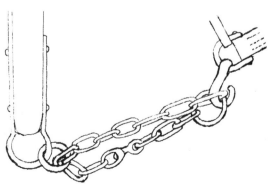

Doubletree with long stub for use in three abreast evener

Singletree with swivel hitch

Standard doubletree

For comparison purposes, a wooden doubletree design

Offset four abreast

Rolled Steel Eveners

Center fire six abreast

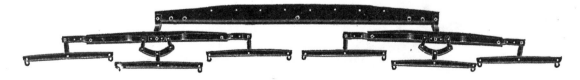

Scissors action center fire three abreast evener

Two styles of center fire three abreast

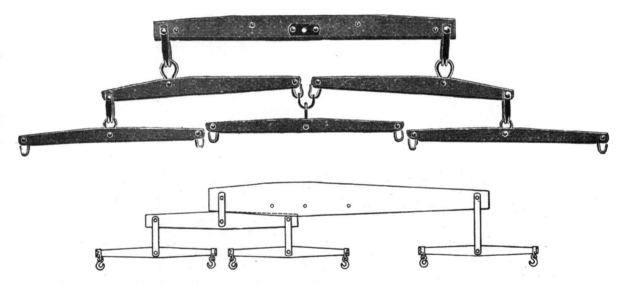

Offset three abreast evener

Legendary Oregon teamster, Morris Elverud, with three young Belgians.

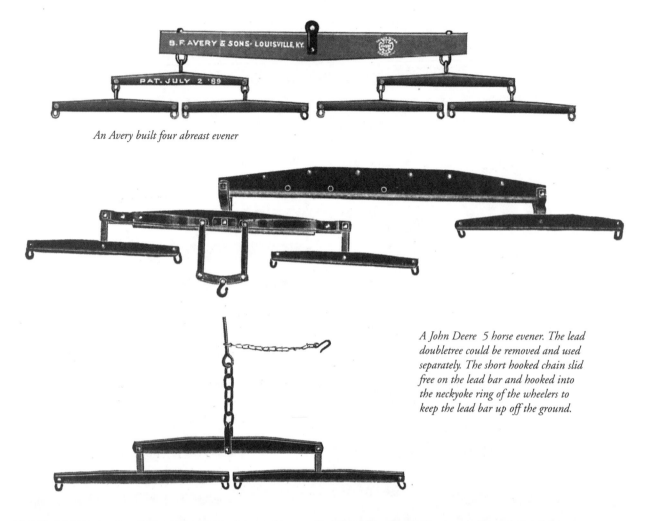

An Avery built four abreast evener

A John Deere 5 horse evener. The lead doubletree could be removed and used separately. The short hooked chain slid free on the lead bar and hooked into the neckyoke ring of the wheelers to keep the lead bar up off the ground.

A beautiful Belgian four abreast belonging to North Dakota horseman, Tom Odegaard. Photo by Fuller Sheldon

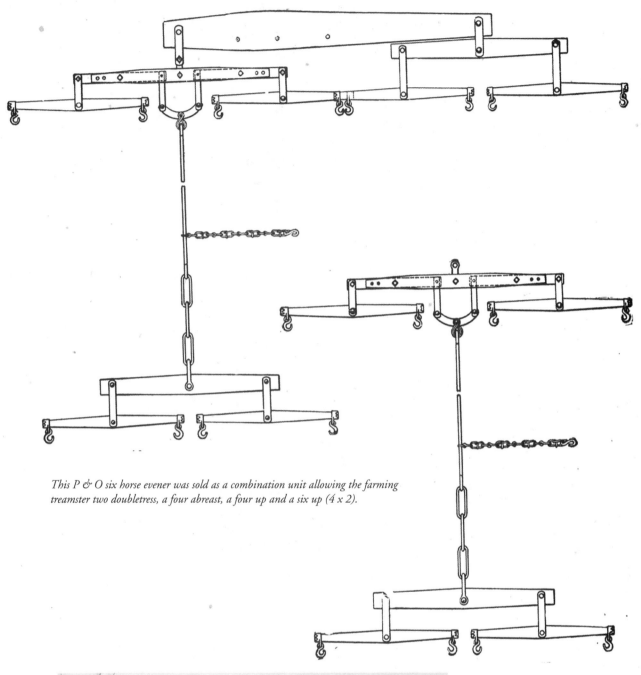

This P & O six horse evener was sold as a combination unit allowing the farming treamster two doubletress, a four abreast, a four up and a six up (4 x 2).

The late Allan Conder drives six of his ponies abreast pulling a spike tooth harrow setup.

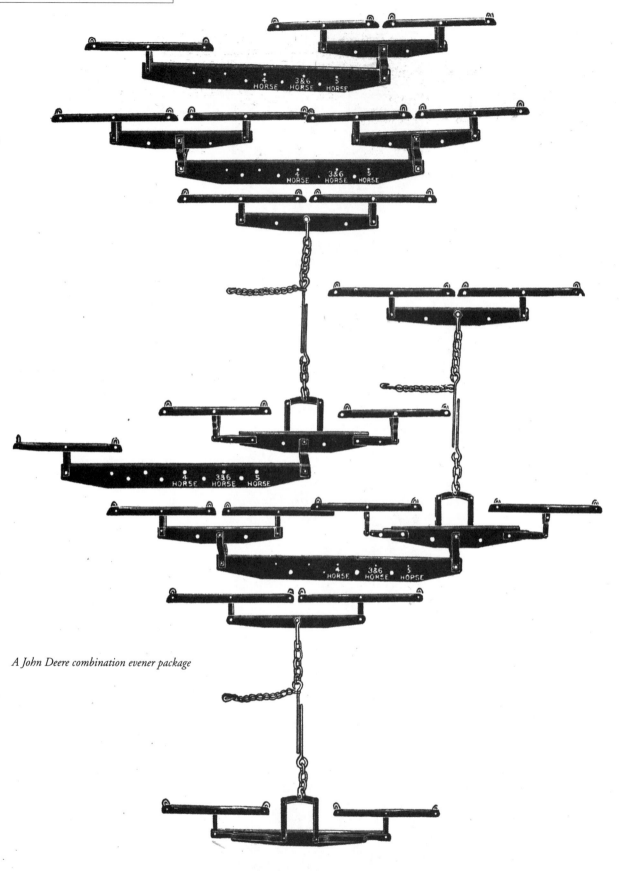

A John Deere combination evener package

No. 61—Six-Horse
No. 62—Five-Horse } Steel Combination
No. 63—Four-Horse-Abreast
No. 64—Four-Horse-Strung-Out

Link Longstaff planting speltz with Ralph, Bart, Bill and Dan in Lancaster PA, Photo by Andrea Longstaff

*The author, Lynn Miller, with Belgian mares Cali and Lana hitched to a John Deere Buck Rake. For specific information on this unique apparatus, including evener displacement and hitching, see the book **Hayng with Horses** by L.R. Miller.*

Additional Information

Though we have offered dozens of variables here, under *hitch gear*, hundreds upon hundreds of evener variables exist in modern manufacture, modern farmshop construction, recent history and beyond. There isn't enough room, even in this large text, to cover it all.

Perhaps a few added miscellaneous notes are required.

New Pioneer Equipment steel four abreast evener.

New Stuff

Several companies are making and selling excellent eveners (see resource directory back of this volume) out of both steel and wood guaranteeing a ready supply. The pipe steel and hardwood eveners are quite heavy for larger hitches used on a daily basis (not so much for the animals as for us farmers). New rolled flat steel eveners identical to the earlier factory production are being made and will grow in favor for their portability.

New Old Stock

Occasionally you may find old wooden factory eveners which look like they've never been used, original paint, stenciled names and numbers, decals, real sharp. Be wary, the eveners themselves may be weak. Depending on where and how they've been stored, the wood, especially in and behind metal, may be brittle and powdery. If you are determined to put them to use, get a leather awl or icepick and push in and around the wood to help determine whether or not the material is solid. If it is, then mix up one part turpentine with three parts linseed oil, warm it, and paint it on the wood, allowing to soak in good. This treatment will make it look real pretty and should help to restore some elasticity to the wood. Best bet would be to display the old wood eveners and get steel or new wood ones.

Paul Reno of Oakland, CA., along with Gene Hilty and George Cabrals, ramrodded the re-creation of a 21 mule "Schandoney" hitch in 2001. The 21 mules pulled 3 separate gang plows with 13 bottoms cutting 143" of furrow.

Rope and Pulley Evener Systems

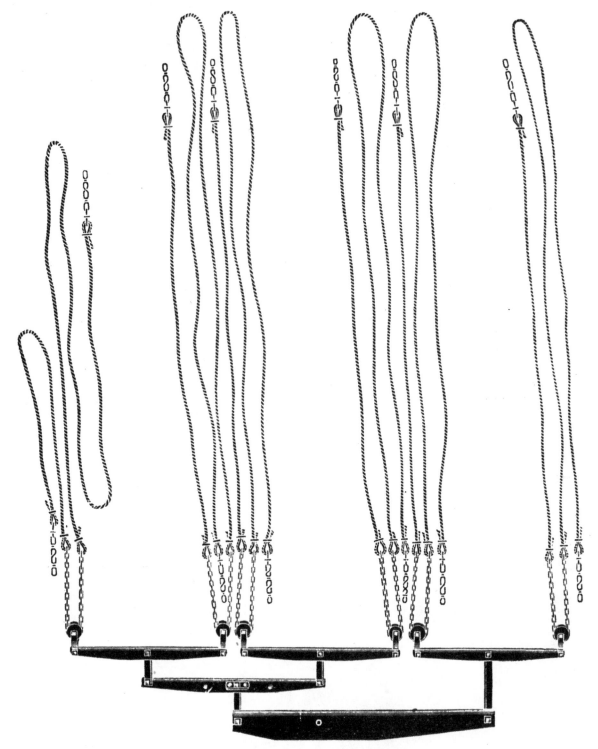

Contrary to what some people think, the pulley setup for multiple hitches is not a new idea. Here we offer a diagram of a six up unit offered by John Deere in 1910. As with all such complex ideas, this system has evolved through several mutations. For an interesting look at a few, get a copy of **Big Teams in Montana** reprinted by SFJ and listed at the rear of this book.

The basic principle at work in this hitch is that one traditional evener is used for the wheelers which has pulleys attached to the ends of the singletrees. Chains or ropes run from the wheeler's trace back around pulley and forward to the leader's trace as shown in the photos on the next page.

rope & pulley eveners

This unusual multiple hitch evener system didn't just drop out of the sky ten years ago. In one form or another it enjoyed a previous life. In its current manifestation it is a direct descendant of a few different ideas which sprung up right after 1900 to allow larger implements to be pulled by more animals.

The Gilchrist Equalizer

The basic principle is one of parallel tandem (and random - as in three in line) hitchs which equalize each lead animal directly back against the wheeler in line. A traditional evener is then used to make all the separate tandems pull equally against each other.

Hard to say what came first but earliest notes speak of **block and tackle equalizers** for "string" hitches where 2 x 2 x 2 or more are employed. The lead team and the team immediately following work against each other, while the teams back of them work on blocks and tackles, giving them leverage of three to one, four to one, etc., against the horses ahead.

A **Schandoney pulley combination** saw use in large multiple hitches. The typical clover leaf, shown in the *RIGGING THE HITCH* pictures, had a chain running forward to a block and tackle running from swing to leaders.

Extra strong **tandem chain and pulley singletrees** were sold years ago. At each end of the singletree a grooved pulley was attached. Through this was passed a short link chain 10 foot 3 inches long including hooks on the end. The ends were attached to trace ends of the two horses which stood one behind the other. The chain had a ring welded in at 29 inches from the rear

end. This was done to prevent the rear or wheel horses from getting too far ahead.

The next derivation appears to be what was called the **"Gilchrist" Equalizer** which replaced chain with rope and attached from hame to hame, doing away with traditional hames (see picture above).

Other mutations of the same idea have been experimented with over the years.

Today, an Amish equipment company out of Gap, PA, (see resource directory) White Horse Equipment, builds and sells a simple rope and pulley evener system. When I spoke with one of their representatives in July of 2003 he said they had sold in excess of 10,000 of these units in the last few years. Another manufacturer, Pioneer Equipment of Dalton, Ohio also makes and sells a rope and pully equalizer.

With the White Horse and Pioneer Rope and Pulley, single tree hook irons are knotted into the ends of the nylon rope which passes through the modern pulleys. The hooks are attached directly to trace chains. See *RIGGING THE HITCH* chapter for more particulars.

Showing the rope and pulley evener doing a good job. Photo by Lynn Miller

This photo shows how the hitch hangs when there is no draft. Photo by Lynn Miller.

In this photo we demonstrate how the short 'carrier chains', which hold up the ropes need to be adjusted to allow an unbroken line of draft when pulling hard. Otherwise there is great pressure down on the wheeler's collars. Photo by Lynn Miller.

North Dakota's Tom Odegaard drives his eight Belgians on a three bottom plow utilizing the rope and pulley evener system. Photo by Fuller Sheldon.

3 abreast offset

The single most often asked question of the first edition Work Horse Handbook was and is what are the dimensions for a 3 abreast offset evener, the kind which allows the farmer to put two animals on one side of the tongue and one on the other while, all three are pulling equal. So I've decided to publish in this edition every single variation we have in our archives.

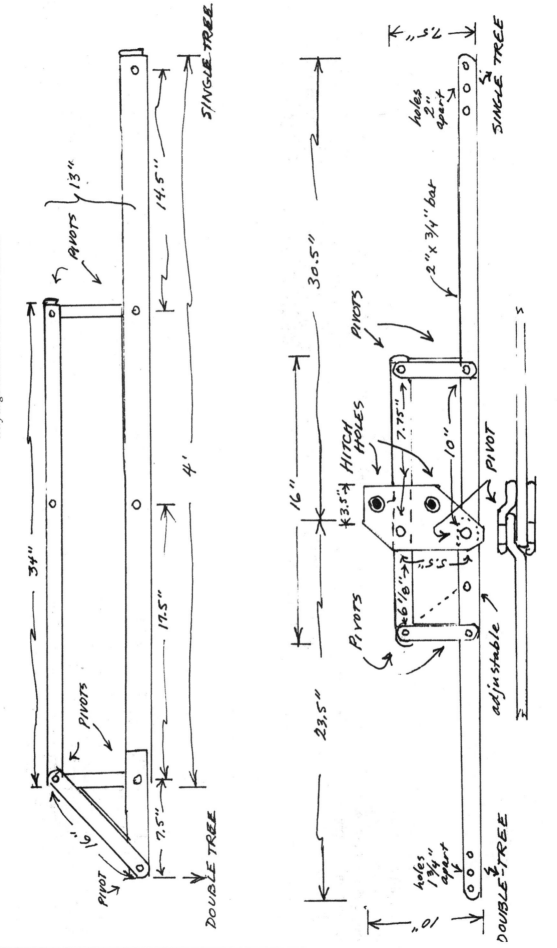

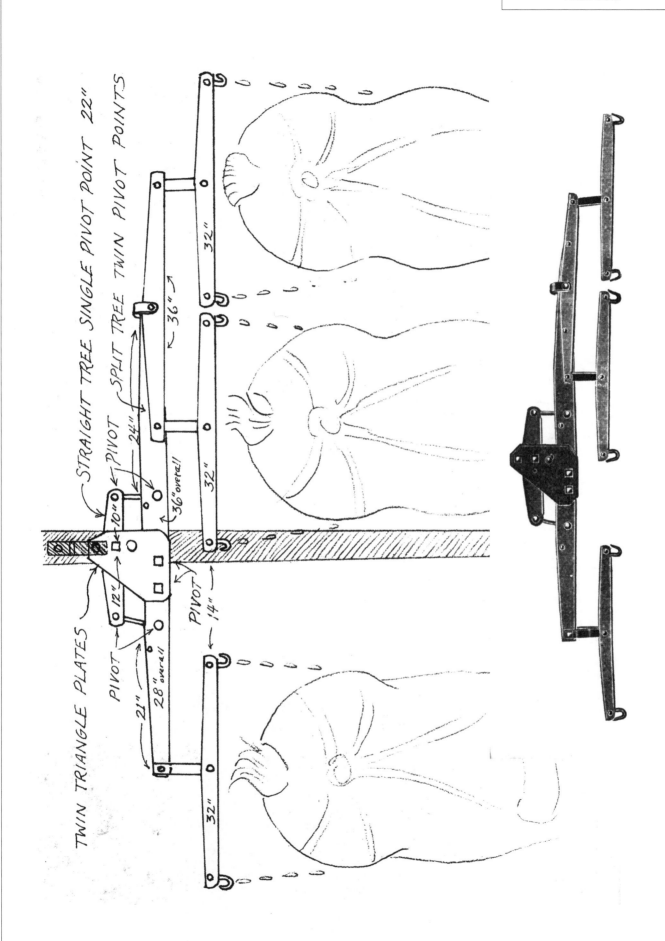

TWIN TRIANGLE PLATES

STRAIGHT TREE SINGLE PIVOT POINT 22"

SPLIT TREE TWIN PIVOT POINTS

PIVOT

PIVOT

PIVOT

PIVOT

10"

12"

14"

21"

24"

28" overall

32"

32"

32"

36"

36" overall

This evener was used with a seed drill.

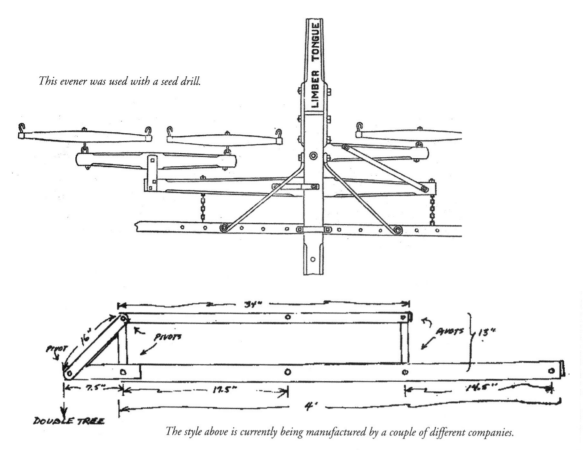

The style above is currently being manufactured by a couple of different companies.

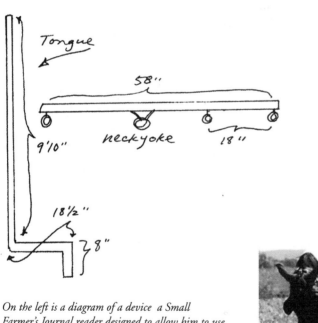

The three abreast is a most popular farm hitch. These Spotted Drafts below were demonstrating at the PA Horse Progress Days of 2000. Photo by Lynn Miller.

On the left is a diagram of a device a Small Farmer's Journal reader designed to allow him to use a three abreast of his ponies on a wagon. The offset tongue was slid into a stub receiver on the wagon.

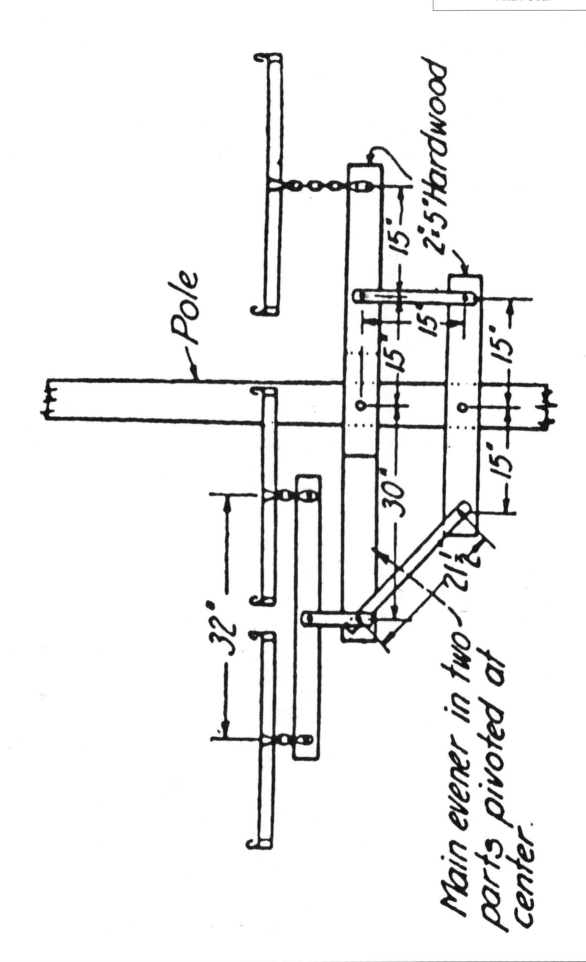

Pole

2ꞏ5ʺHardwood

15ʺ

15ʺ

15ʺ

15ʺ

15ʺ

15ʺ

30ʺ

21ʹ

32ʺ

Main evener in two parts pivoted at center.

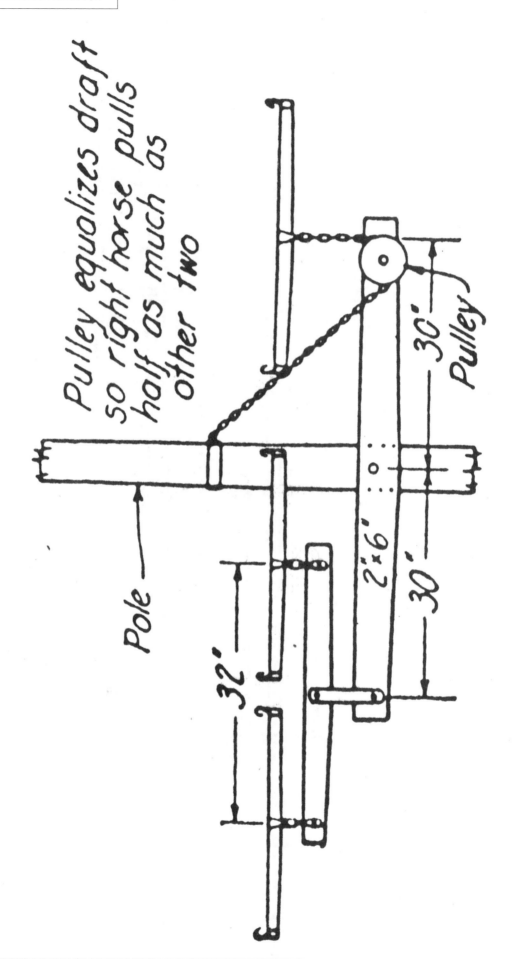

Pulley equalizes draft so right horse pulls half as much as other two

Pole

Pulley

30"

30"

2"x6"

32"

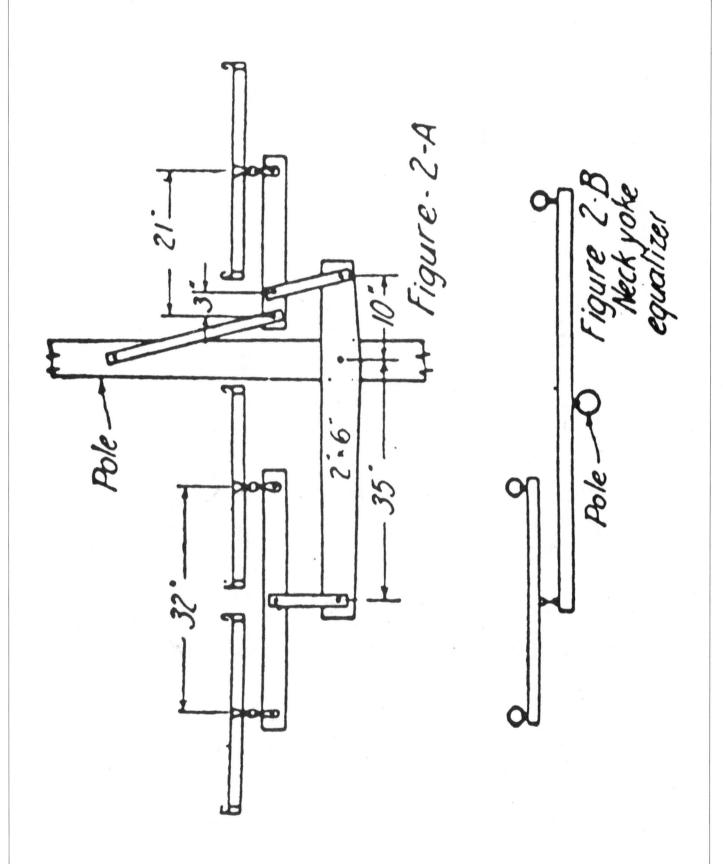

Pole

21"

3"

10"

2'-6"

35"

32"

Figure - 2-A

Figure 2-B
Neck yoke
equalizer

Pole

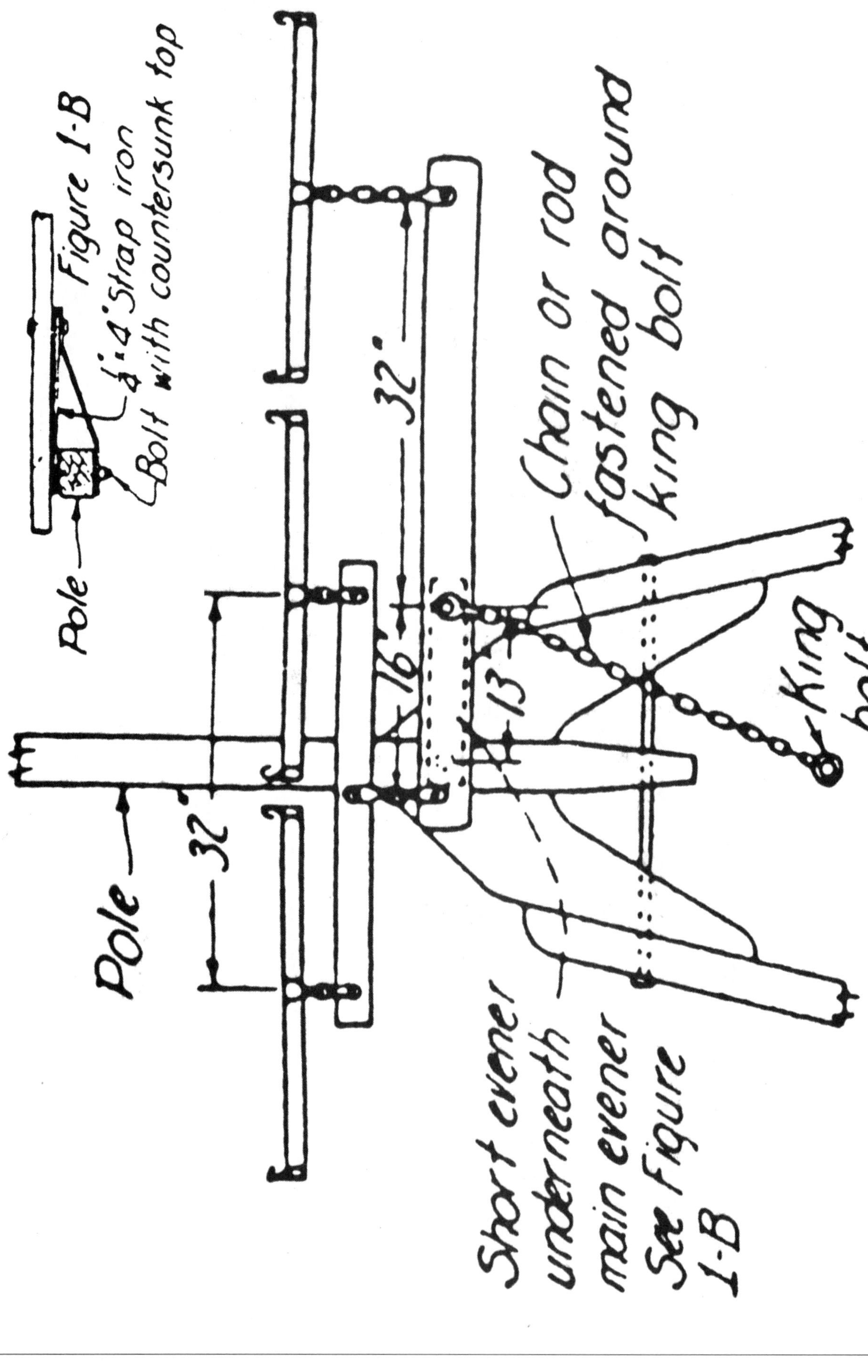

Figure 1-B

Pole

⅜:4" Strap iron
Bolt with countersunk top

Pole

32"

32'

16

13

Pole

Chain or rod
fastened around
king bolt

King
bolt

Short evener
underneath
main evener
See Figure
1-B

Figure 1-A

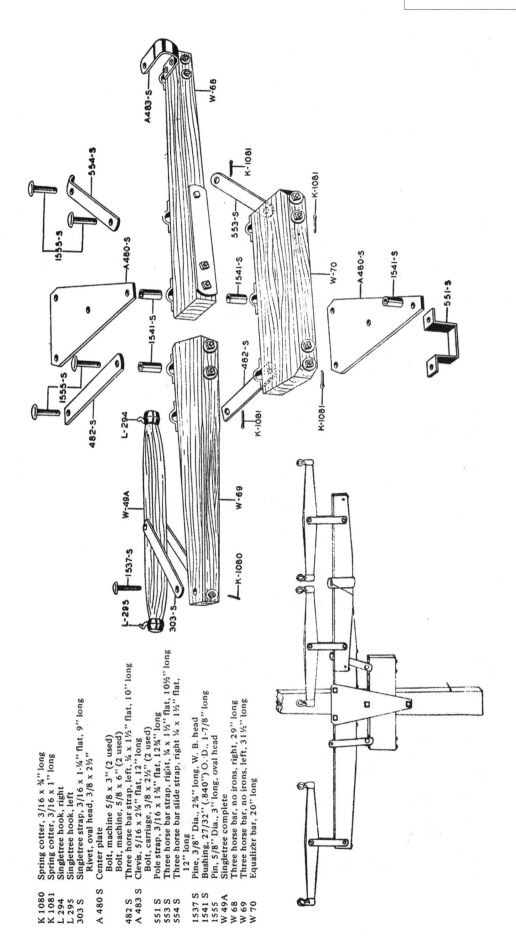

K 1080	Spring cotter, 3/16 x ¾" long
K 1081	Spring cotter, 3/16 x 1" long
L 294	Singletree hook, right
L 295	Singletree hook, left
303 S	Singletree strap, 3/16 x 1-¼" flat, 9" long
	Rivet, oval head, 3/8 x 2½"
A 480 S	Center plate
	Bolt, machine 5/8 x 3" (2 used)
	Bolt, machine, 5/8 x 6" (2 used)
482 S	Three horse bar strap, left, ¼ x 1½" flat, 10" long
A 483 S	Clevis, 5/16 x 2¼" flat, 12" long
	Bolt, carriage, 3/8 x 2½" (2 used)
551 S	Pole strap, 3/16 x 1¾" flat, 12¾" long
553 S	Three horse bar strap, right, ¼ x 1½" flat, 10½" long
554 S	Three horse bar slide strap, right ¼ x 1½" flat, 12" long
1537 S	Pine, 3/8" Dia., 2¾" long, W. B. head
1541 S	Bushing, 27/32" (.840") O.:D., 1-7/8" long
1555	Pin, 5/8" Dia., 3" long, oval head
W 49A	Singletree complete
W 68	Three horse bar, no irons, right, 29" long
W 69	Three horse bar, no irons, left, 31½" long
W 70	Equalizer bar, 20" long

wilbert's rings

On a visit to Missouri back in the early eighties, Wilbert Hilgedick showed me how he took two inch wide leather circles, cut slits to the center of the circle, and cut a larger hole in the middle. These rings were then threaded over the cross checks of team lines as a safety. They worked to keep the splice conway or buckle from passing through the hame ring and hanging up, rendering the lines useless.

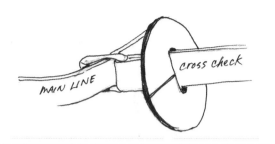

brichen rope

My dear friend, Dr. Doug Hammill, of *Ask a Teamster* fame, uses a brichen rope in training and as a restraint when backing up a ground driven team. The rope snaps into the breast strap hame ring and follows back around the brichen and across to the other horse, continuing on around that brichen and forward to breast strap ring. This places the force of the restraint on the hames and collars and not on a harness part, say the brichen rings, which might break.

jockey sticks

A jockey stick, typically 4 foot long, may be made of wood or steel and features a heavy snap on both ends. Every horsefarmer should have a couple of these handy. They can sure save a day's work. On the front end, they are customarily used to keep an aggressive horse from bothering its team mate. More often than not they are snapped into the opposite side bit ring of the offending horse and back to the other horse's hame. I prefer to snap them into a halter worn under the bridle and in this way avoid abuse to the horse's mouth. Another prevalent use of the jockey stick, seen below, is as a hitching aid for 3, 4 or more abreast. In this case lines are running to the two center horses and jockey sticks force the outside ones to follow.

Photo by Lynn Miller

CHAPTER TEN

PRINCIPLES OF DRIVING FOR WORK

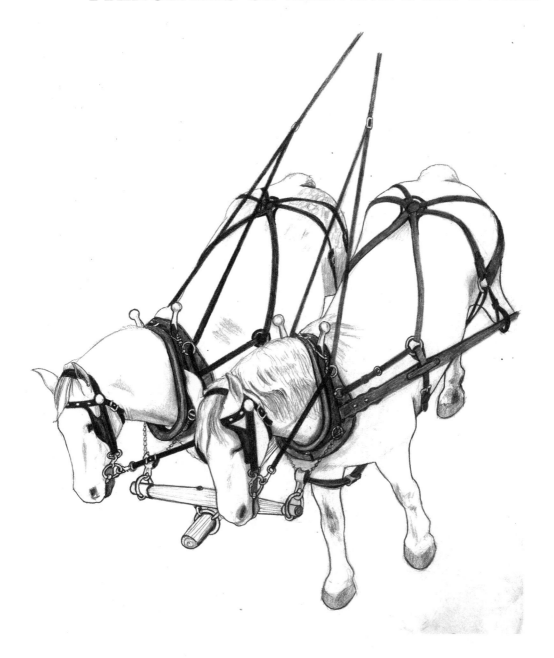

Driving the horse is, in its finest sense, the true reward of understanding, trust and communication between the animal(s) and the teamster. Driving the horse, in its worst sense, may be the awkward outcome of tricking and/or forcing a caged or bound, sometimes terrified, animal into unwanted activity. These two, of course, represent a world of difference and there is a whole range of variations in between. For the beginner must come the realization and understanding that he or she is fully responsible by half (if not more) for this potential relationship with the draft animal. The difficult aspect for most people to accept is that the animal is also responsible by half (if not more) for this potential relationship. It is for this reason that the author makes this strong recommendation:

The beginner should place a premium on those qualities in a work horse which will make a safe and trusting introduction into driving for work. Experience, training, and quiet, docile temperament are all qualities to look for. These qualities should be looked for in the previous handler/owner of the animals as well because this is, or may be, a good indication that the apparent character of the horse runs deep or shallow.

If the animal seems quiet to a fault, with no spunk, and if the teamster brags loudly of his skill and the tricks and abuse he has used – be suspicious. If you know how to translate the statements and pauses, most anyone selling a horse will tell you all you need to know about the animal, even without wanting to.

effect of the beginner

No matter how quiet, willing and well-trained a horse might be (whether old or young), an **inexperi-** enced teamster, with the best of intentions, can ruin the animal in a very short time. Consider this: if you were deep-sea diving for years without mishap and one day almost drown because a newcomer on the boat fouled up the rigging – how would you feel about going right back down again, leaving your life in those same hands again? The same feelings are true for the horse in harness. This is the reason that the beginner cannot afford to jump into driving horses without having an experienced teamster around at first. No amount of information from magazines or books will ever convey to the uninitiated the incredible subtlety and fine balance required of a good teamster. And the enormous trust required. That will only come, slowly, of experience tempered by careful observation.

If a beginner, or interested person, has the opportunity to attend a draft horse driving clinic, school or workshop, it is in their best interests to do so. And a premium should be placed on hands-on instruction (when available), in other words a learning experience where a maximum amount of time with the

I never knew him personally but whenever Fred Baker showed up, back in the 70's at the plowing matches, I watched him like a hawk. When the three of them, man and horses, moved it was hard to tell them apart. Watching Fred was always the best schooling. Photo by Nancy Roberts.

The author driving in 1977. Photo by Scott Duff.

driving lines in hand is offered. It is in such a setting that instruction/guidance will make profound impressions during those critical first moments.

Hopefully your first experiences will be positive (without unnecessary accidents), and you will quickly begin to absorb a reserve of natural responses and reflexes that will serve you in your work with horses. I like to think of this 'reserve' as my vocabulary of response.

Remembering back to this author's first times with driving lines in hand: the first thought is of the awkwardness of the process. That is oversimplifying the experience because so many feelings pass through

simultaneously, such as: the evasive simplicity of the process, clumsiness, power, maybe a hint of terror, a frustrating feeling of being totally lost – so many feelings and they all serve to confuse the learning process.

empathy helps

The only way to successfully communicate how to drive horses is to reverse the picture and try to convey what the animal goes through, feels, sees, and is asked to respond to.

Try to put yourself in the animal's position. First of all, realize that you do not speak the same

Dave McCoy coaxes his logging team through a log skid competition at the Draft Horse Festival in 1980, Eugene, Oregon. Photo by Nancy Roberts.

Oregonian, Dale Greenough, speaks to his horses through the lines circa 1979. Photo by Nancy Roberts.

Rocky Mountain teamster, veterinarian, storyteller and draft horse instructor extraordinaire, Doug Hammill, with two of his Clydedales on a dump rake. Photo by Laurie Hammill.

language as this stranger you must deal with. You are in a leather suit of straps with a piece of cold steel in your mouth, and probably, leather flaps restricting your vision. The steel in your mouth seems to have differing pressure, a pressure which apparently corresponds to the two leather straps connected to either end of the steel (bit). You get an array of signals; confusing voice commands, maybe a slap of the leather straps (lines), maybe some pressure on both ends of the bit – maybe one side, maybe none. From all of this you somehow figure out that a forward motion is expected. Now, back up a minute . . . .

If you, as a so-called thinking human, actually had to go through such a process would you:

1. Be "willing" to cooperate;

2. Be able to "figure out" (reason) what is expected of you?

This author continues to be amazed at the depth of the horse's inherent intelligence, willingness and ability. (And equally amazed at the inability and shortsightedness of many humans when it comes to their relationship with horses.)

be sensitive - think expect cooperation

perfect tension

Imagine (before doing) that the horse has a bit in its mouth and you have two leather ribbons (lines), one in each hand which attach directly to the sides of the bit. Obviously, if you pull back with both hands you apply equal pressure by way of the bit to the horse's mouth. If you pull back with the right line you will be pulling on the right side of the bit and vice versa for the left side. If you slack up on the lines the bit should ride loose in the horse's mouth. So the horse's mouth is working as a signal point, relaying to the animal, from your hands – through the lines – by way of the bit, a signal that should result in a desired maneuver.

The horse's mouth is sensitive, just like yours.

If the animal's mouth is required to take a great deal of abusive, hard, jerking pressure from the teamster it will result initially in sores and cuts and ultimately in calluses and desensitized tissue. So, in time, a horse will be incapable of quick fluid reaction to subtle pressures. The horse will become "hardmouthed."

If there is no pressure from the lines, the horse will be confused about desired maneuvers and a mess will result. Loose lines also make it difficult for the teamster to respond quickly, if necessary, to problems. And third, loose lines stand a good chance of getting tangled in some other part of the harness. The teamster, then, must find a delicate balance of just enough line pressure if optimum performance is the goal. And to achieve that "perfect tension" when actually driving is a difficult task as each change in motion and direction requires a sensitive give and take with the lines. As the animal moves, to maintain an always even tension, without abusing the horse's mouth, requires that the teamster use the arms like soft shock absorbers or springs or hydraulic cylinders always moving to find balance.

To illustrate what we mean, we'll go through the steps of ground driving a single horse and then ground driving a team of horses.

Note: This is a complex topic, one which we have devoted another entire book to. What follows is a basic introduction. If you are serious, we recommend that other book, **Training Workhorses/Training Teamsters** *(see resourec directory this volume), and that you put your feelers out for good individual help.*

My dear friend, the incomparable late Ray Drongesen, with King and Ruby mowing hay with his Champion on my Junction City farm in 1975. Ray always made it look so easy, and it was for him. Pure pleasure and all comfort. Photo by Lynn Miller

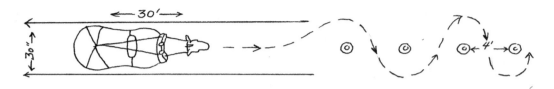

A simple single horse driving excercise.

ground driving
the single horse

Use a well-schooled and experienced horse and have a teamster handy to help. Since this is an exercise and no vehicles or load will be pulled or drawn, the type of harness involved does not matter much. With a bridle and bit in place and with collar and hames set up properly (see *Harnessing* chapter), snap or buckle one single line to the right side bit ring and pass the line through the top right side hame ring and on back. Do the same with the left side. With the right line in the right hand and the left line in the left hand, stand well back of the horse and take up the lines so that there is only a little slack (NEVER allow a big belly in the lines). Give a verbal command to go. You should know what command the horse is accustomed to. (Do not slap the horse with the lines unless as a last resort to prevent balking.) As the horse steps ahead, find an even tension for both lines and maintain it. Be careful, however, not to hold the lines too short or pull back too soon as the horse first steps ahead because you could stop him before he gets started and further confuse signals later. So, as the horse walks, practice pulling in a little with one line while letting the other line out an equal amount. Watch how the horse turns to follow its head. Now pull back both lines steadily and say "whoa" in a clear, firm voice (no need to shout). The horse should stop. When the horse stops, let up on the pressure to both lines. To make the horse back, pull on both lines evenly and say "back." Hopefully, you will understand why to slack up on the lines after the animal has stopped. The novice will often end up stopping a horse only to find the animal backing up. If you want to stop, just remember to slack up on the lines after the animal quits its forward motion (at the whoa command).

Now back to the driving. Try this exercise. With an even tension (no slack) give the command to go and stay behind the animal but moving slightly to the right. Do not "steer," just walk, maintaining an even firm tension on the lines and passing, while behind the animal, to the right side. The horse should be making a slow, steady left turn as you stay behind the animal (moving right). This exercise should give you the physical sense of how subtle the turning process actually is. It should not be necessary to yank an animal over to make a gradual turn.

Another exercise is to pull the left line slightly and let out an equal amount of right line – the animal should turn left. Then try a right turn. Now all this may seem too basic. The real test follows. Draw or scratch or otherwise mark two lines 30 inches apart and 30 feet long. Then in another area set 4 pylons 4 feet apart. Now ground drive the single horse "straight" down between the two lines and then weave the pylons (as illustrated), driving the animal exactly where you want to go. It will sound simple. It will BE difficult the first time. You will probably hold one or both of the lines too tight or too loose and the horse won't go where you want to go. Don't blame the horse; he's only following your instruction (or making his judgment of a mass of confusing signals). If things don't work right, stop and try to understand what you're doing wrong. Practice ground driving until you at least have some measure of respect for what you don't know and an appreciation for how much understanding the horse is capable of.

ground driving the team

With an experienced teamster plus two well-schooled, harnessed horses accustomed to working together, set up the driving lines as explained and illustrated in *Rigging the Hitch* (this text). Take up the lines, left in left hand – right in right hand, and step well back of the team. Take up any slack in the lines and give a verbal command to go, allowing the lines to come to a perfect tension in your hands. As the horses move ahead, follow them and just try to maintain

Above and below: Anita and Ron VanGrunsven, husband and wife horsefarmers. These folks farmed a big market garden near Portland, Oregon and recently moved to Idaho.

(Right) The author driving a team at his first Horsedrawn Auction in Albany, Oregon in 1979. Photo by Nancy Roberts.

Right: The legendary Dan Kintz showing the style, horsemanship and quiet manner which won him so many pulling matches across the northwest. Photo by Nancy Roberts.

(A.) These photos demonstrate a simple cross-over maneuver for gathering in lines. When one line is slack...

(B.)...reach across, with other line in hand, and take hold of both lines.

(C.) Now pull up the slack, with the one hand holding both lines...

(D.) ...and take hold of the lines, one to each hand.

(E.) Or you can simply cross over and pull the slack through your hand.

perfect tension and keep the team moving straight as an arrow. Then pull both lines back steady and say "whoa." The team should stop and you should slack off on the lines. With even tension on the lines, ask the team to "back." Then ask the team to go ahead again, this time carefully navigating a right turn. Pull in on the right line and slack up the same amount on the left line. Notice that the right lines go to the right side of the bits on both horses (and vice versa left). Now try a left turn. As you drive, practice keeping your arms out ahead of you in a comfortable position. This allows room for you to stop or turn the horses quickly if necessary. If your arms are tight against your body, incapable of movement, and there is slack in the lines, YOU ARE COMPLETELY OUT OF CONTROL, NO MATTER HOW THE HORSES ARE PERFORMING!

The novice will experience a hesitation about allowing the lines to pass through the hands or in taking in line. Page 216 illustrates a simple maneuver for taking in slack while making a turn. It may seem unnecessary to explain such a simple thing, but when the novice, no matter how high the IQ, is lost in a panic of a new moment, the most basic of instructions is vital.

It takes only a moment for the beginner to feel the difference between the single horse and the team. Hopefully, the beginner will acquire a special respect for those reasons why first experiences should be had in the company of a skilled teamster.

commands

We've spoken of giving the command to go. You will develop a habit for that sound or word or phrase which is most comfortable. Suggestions include a short whistle, a kissing noise exaggerated, a clicking and/or clucking noise ("giddeeup" seldom works unless the individual animal has been used to it).

It is a good habit to say "Whoa-BACK" to standing

Willard Wilder skillfully guides six Percherons in a 1980 Oregon Draft Horse Festival show entry with Iowa auctioneer, Bill Dean, riding shotgun.

animals for a backing command as they may not hear the simple "back" And FEEL the tension on the lines and get a crossed signal. For this reason, make sure that the animals are listening: LOOK AT THE EARS!

"GEE" (as in gee whiz) is the universal English-American command for right turn. "HAW" is the command for left turn.

It is always a good idea to speak to the animals by name often. Before giving any command, WATCH THE HORSE'S EARS and MAKE CERTAIN ALL ANIMALS ARE LISTENING and EARS ARE FACING BACK.

After you work horses for a few years, you might be pleasantly surprised to find that they can understand and follow a variety of simple verbal commands.

hands holding lines

We spoke early in this chapter about the need to be sensitive about the bit in the animal's mouth and how you apply pressure. If you listen to teamsters and

Stephen Decater, of Covelo, California. Photo by Nancy Warner.

Holding lines over the fingers and under the thumb. This position is a second choice as it is moderately sensitive.

Holding the lines between the first two fingers. This is one of the first choice positions (along with the one below) as the teamster "feels" in his hands more closely an approximation of what the bit "feels" like in the horse's mouth.

Holding lines under the hand. This position is the least sensitive, or emphathetic with the horse, and is used where the greatest leverage or force is needed.

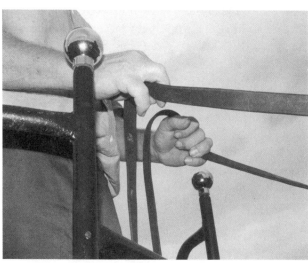

Lines between third and fourth finger, a good position.

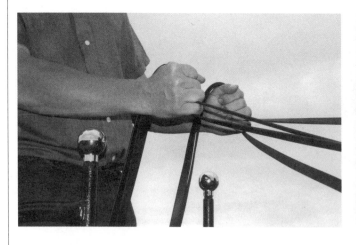

One way of holding the lines for two spans.

In this picture the six lines control twelve horses in three spans of four abreast.

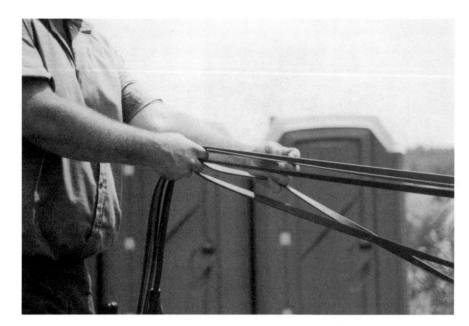

Showing six lines in hand (for the twelve horse hitch) and the pull required for a slight turning action. Notice, below, the slack in the lowest near line. The lead and swing team are being pulled into a turn and that slack line, to the wheelers, is allowing them to continue for a bit straight ahead.

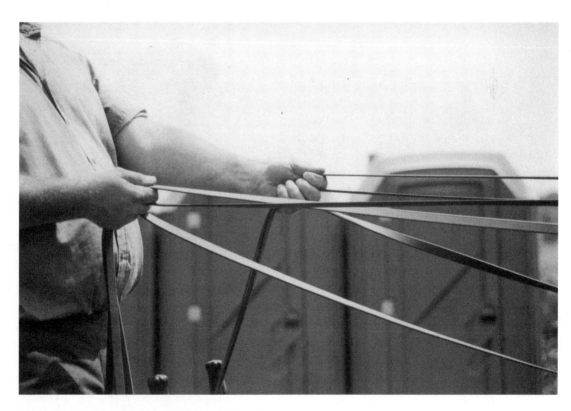

read what literature is available, you will find passing (and reverent) reference to old so-and-so's "hands." It is a high compliment to say a teamster has good hands. The best teamsters control hitches of any size with the slightest of pressure and a soft, sure voice. It is a finesse. But to say a teamster has "good hands" is misleading. What is meant ultimately is that the good teamster is a sensitive human who is always culturing that sensitivity and never taking certain procedures for granted.

When driving with two lines, whether it be a single horse, two or whatever, there are three natural positions for holding lines (see page 218); over the hand, under the hand and between the first two fingers. The latter will more closely approximate, in the teamster's hand, what the bit feels like in the horse's mouth. And it is just such an "earned" empathy which

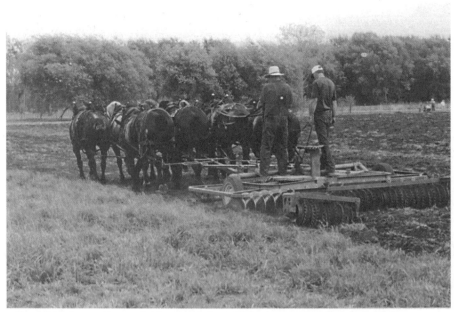

Iowa's Dick Brown discing a field with eight beautiful Percherons. Photo by Bob Mischka.

serves to build that particular sensitivity of which we speak.

As we said at the beginning of this chapter, driving can and should be the reward of a special kind of teamwork between teamster and animal(s). But that's up to you. If you find you must force your way in this craft, you will tire of it quickly and never appreciate the possibilities and the beauty.

The TV western notion of slapping standing horses with the lines and hollering to get them to go is an insensitive and exaggerated minor procedural myth. It is a bad habit for beginning teamsters and should be avoided with a fervor. Why? Well, imagine what you are doing by conditioning the animal to go whenever it feels a slap on the back and/or a holler. There are any number of possible situations where an unwanted "signal" to go will fall on them, whether it is a branch in the woods or a neighbor's pat of the hand. It is infinitely better to develop a full vocabulary of easily understood words and to trust the animal's great ability to understand them and respond to them. Also remember that the horse has an acute sense of hearing, so you need not yell all the while. As you develop your vocabulary with the horse, you will also be developing a unique set of "keys" to your power system.

The author getting to know a young Belgian mare Photo by Renee Russell.

CHAPTER ELEVEN
RIGGING THE HITCH

A grade draft horse hitched single to a cart using a basic brichen farm harness.
Ray Drongesen driving Ruby with Juliet and Ian Miller riding. Photo by Christene George

This chapter illustrates the configurations and positioning of fourteen different hitches. Actual hitching-up procedures are covered in the next chapter. The information in this chapter should serve also to impart a sense of the great flexibility of true horse power. The teamster, for instance, with four horses, has the possibility of four single horses each hitched separately, one single and a unicorn hitch, two singles plus a team, or two teams. In other words, with additional teamsters to drive, the four horses can be hitched more than twelve ways, culminating in four abreast or four-up. Try to imagine taking one large tractor and dividing it into four little ones as the need arises.

In North America, the team (two horses hitched abreast), is the most popular hitch with the single being a long second. In the British Isles, the opposite is true, with singles being the most popular hitch. With so many small farms taking to the use of one horse or mule, there may be a change in view in North America.

the single horse

Without going into all the possible derivations in single horse harness, let's just say that the same basic harness used for teams will work for single with minor modification. First of all, if the horse will be drawing something other than a wheeled vehicle, it will probably not need a braking system to the harness. A cruper harness would suffice for work such as cultivating, plowing, logging, etc. A brichen harness would be necessary for wagon or cart work or anything which required that the load be prevented from rolling or skidding upon the horse.

A cruper (plow) harness (see page 261) consists of

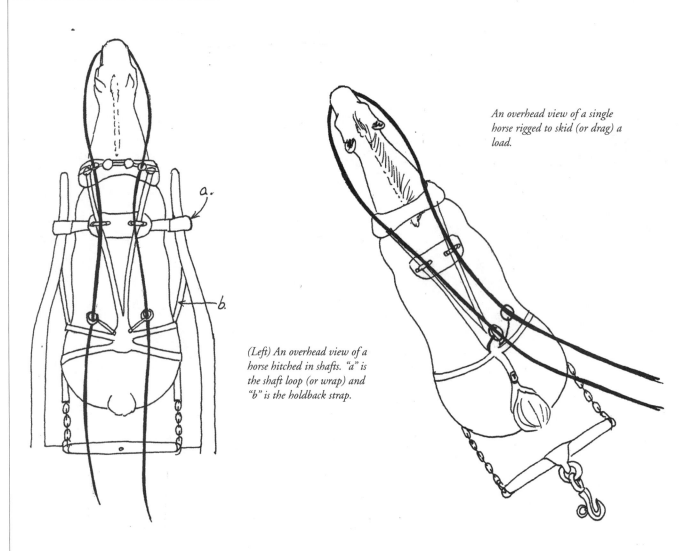

An overhead view of a single horse rigged to skid (or drag) a load.

(Left) An overhead view of a horse hitched in shafts. "a" is the shaft loop (or wrap) and "b" is the holdback strap.

all the basic parts with the exception of the brichen/quarter strap assembly. "Cruper" refers to the strap which fastens around the tail head to secure the harness from sliding off to one side or the other. A brichen harness (see page 111) includes the brichen/quarter strap assembly which functions (whether used team or single) as the braking and backing system.

The driving lines for a single horse consist of two single leather lines, 3/4" to 1 and 1/4" in width and anywhere from 18' to 24' in length, with either snaps or buckles at one end of each line. We recommend lines with buckle billets. The bridle and hames used for a single horse are the same as that used for a team. I recommend that you have two spreader-type straps with a ring in one end and a snap at the other end. The strap should be anywhere from 4" to 14" in length. I call these "line keepers."

Whether you use a cruper or a brichen harness, the lines are set up the same. I should mention that a brichen style harness can be used for all types of work,

whereas the cruper style has its limitations. Also in a brichen style team harness, a breast strap and a pole strap are used for the neck yoke assembly. In single horse work these parts are not necessary. If the horse is to be used to pull a two wheeled cart with a heavy load front of the axle center, then it will be important to make sure that the back pad is sufficiently large and clean of sharp or uncomfortable edges. A little padding in the back pad might even help. The back pad should sit right where a saddle normally would, directly behind the withers of the horse. The reason for all this concern is that with the horse between the shafts, all the front weight of a two wheeled vehicle is pushing square down upon the back pad, the weight being transferred down the shafts through the shaft loops to the back pad.

In North America, teams of horses are hitched to wheeled vehicles with a single tongue or pole extending forward from the vehicle and between the two horses. This tongue or pole is attached to the

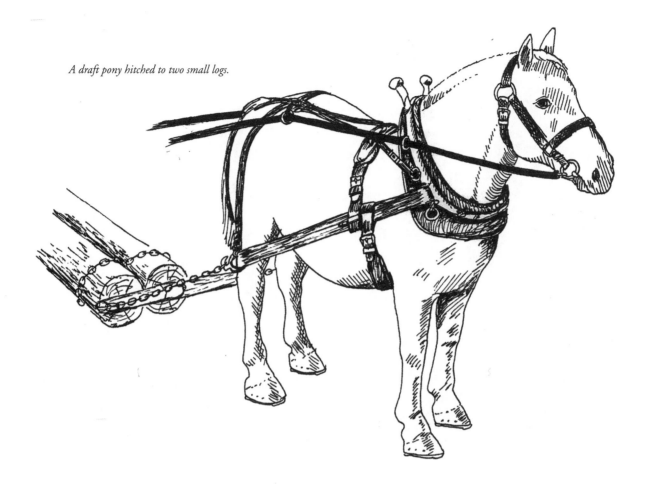

A draft pony hitched to two small logs.

horse's braking and backing system usually by way of a neck yoke. When hitching a single horse to a wheeled vehicle, the horse is placed between two poles, or what are referred to here as shafts, which extend from the front axle forward. As with the wagon tongue, when the shafts are turned from side to side, the vehicle also turns. To be of proper length, the shafts should not extend beyond the front of the horse when hitched. Obviously, there is great variation in length of shafts depending on the size of horse used. Same thing is true of the distance between the shafts. Most buggies are set up to be pulled by 1000 lb. trotting horses and a 2000 lb. draft horse won't fit between the shafts. Modifications can be made to allow that bigger or smaller horses be hitched to any vehicle. (See page 185.)

The harness required for a horse between shafts can be especially designed for the purpose, fancy or plain, or it can simply be a farm brichen harness with a few extra straps. The basic hitching principle remains the same. The shafts, one on either side of the horse, are held up either by passing through leather shaft loops which attach down from the back pad and over the tug, or can be strapped in the same position

with two hame straps attached to the "D" rings at the end of the belly band and back pad billets (see page 225). Next, "holdback" straps are attached to the end rings of the brichen. These serve as the braking and backing system. The holdback strap is attached to the shaft just back of where the "shaft loops" go. The attachment usually consists of a metal loop fastened to the underside of the shaft. An adjustable strap, about the length of a harness quarter strap, passes through the loop and around the shaft. On the end of the strap is a heavy snap which hooks to the brichen ring. (See page 225.)

With the shaft loops and holdback straps fastened, the tugs pass over the holdbacks to the singletree. With tugs and holdbacks fastened you want just a little slack, not much, in both. It will take a little doing for the inexperienced teamster to figure out proper adjustment of the straps. Here again, it would save time and grief to start out not only with an experienced horse but also with a harness and vehicle already set up and properly adjusted. With a four-wheeled vehicle there is no significant weight passed through the shafts to the horse's back. With a cart or

(Above) A photo of a French-Canadian (or Canadien) draft stallion hitched, in breast collar harness, to a buggy at the Upper Canada Village in Ontario. Photo by Lynn Miller

(Below) A Percheron working a horse power (gear drive) which powers a drag saw for cutting cord wood. Photo by Lynn Miller.

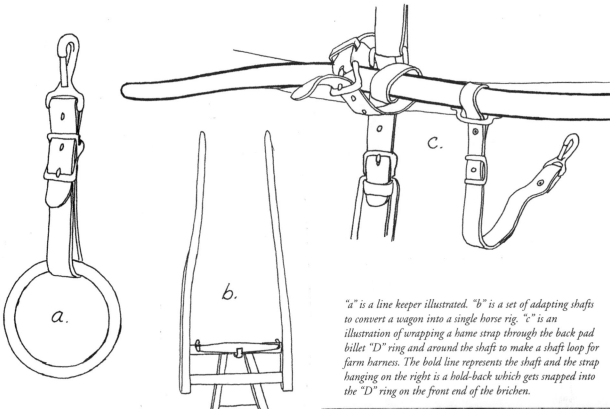

"a" is a line keeper illustrated. "b" is a set of adapting shafts to convert a wagon into a single horse rig. "c" is an illustration of wrapping a hame strap through the back pad billet "D" ring and around the shaft to make a shaft loop for farm harness. The bold line represents the shaft and the strap hanging on the right is a hold-back which gets snapped into the "D" ring on the front end of the brichen.

two wheeled vehicle, any weight forward of the axle can be expected to apply pressure to the horse's back. With a healthy, big horse and proper harness, up to 200 lbs. of actual weight is no problem. Yet it would save the horse energy if the weight could be balanced over the axle with only a slight extra ahead of the axle.

Eric Nordell, Pennsylvania horsefarmer let's his walking plow team take a break. Eric and Anne Nordell are regular contributors to SFJ.

A scene in the Millersburg, Ohio area. Amish cart pulled by single pony. Photo by Kristi Gilman-Miller

Bob Nygren with Lucky and Jewell in 1978.

the team

The term 'team,' when used to describe working horses, commonly means two animals 'hitched' side by side (or abreast). In some regions, 'team' means any 'hitch' of two or more animals working together. In the southern United States, it is common to hear people refer to a 'span' of mules (or even horses). Here again, depending on the particular region, span might mean two animals or it might be inclusive of two or more animals. For the purposes of simplicity and clarification, this text refers to team as two animals working side by side (as illustrated here) and leaves the definition of 'span' to future campfires.

In chapter seven, HARNESS, the parts of the harness were described and mention was made of the function of team harness parts. In chapter eight, HARNESSING, illustrations indicated how to harness a horse with a team brichen harness. In chapter nine, doubletrees and neck yokes were described and illustrated. And in chapter ten, DRIVING, you were taken through the exercise of driving a team. Hopefully, the illustrations you see here will cement all that information into a solid picture of the structure and dynamics of two horses working together.

First, a quick review of the basic difference between harness used on a tongue (for a wheeled vehicle), and harness used to skid a load: When a team, or any hitch, is hooked up to a wagon or implement which rolls free, something must be done to restrict the movement of that vehicle. Yes, the first concern might

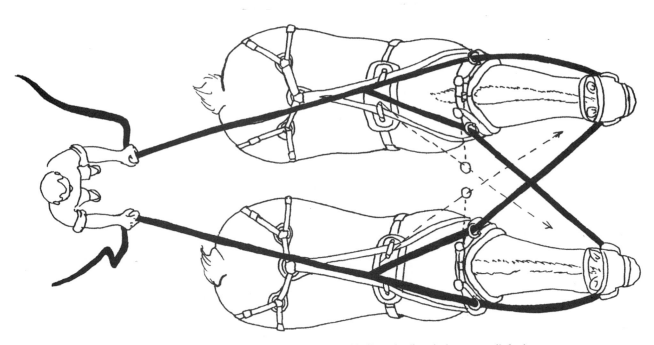

Team lines threaded, dotted lines represents spreader use which would effectively allow the horses to walk further apart.

be how to hook on and pull the vehicle, but of equal concern must be the restraint, when necessary, of movement. Or in other words, backing and stopping the load.

The harness for hitching to a tongue must include, at the very least, a breast strap. This author recommends the use of a full brichen harness including quarter straps, pole straps, and breast straps as illustrated in Chapter Seven. Page 228 in this chapter illustrates how this assembly hooks to the neck yoke which in turn fastens to the tongue.

When skidding a load, where there is no concern that the load might run up on the animals, no brichen is necessary and a cruper or 'plow' harness is adequate. There is no reason why a brichen harness, however, shouldn't be used to skid loads as well.

Lines. The driving lines used for a team are quite different from those used for a single horse. There are illustrations, in Chapter Seven and above. which illustrate how the lines are used. As stated before, in driving horses, the lines are your critical contact with the animals. Never use weak, dry or poorly constructed lines; it isn't worth the possible trouble.

With team lines, the cross check is normally a few inches longer than the main line. The main line, usually a continuous (spliced) line of leather, always

A team of Belgian mares belonging to the author.

This drawing illustrates the neck yoke fastened to the breast straps and the end of the tongue. Note also the line setup with spreaders and in the middle, the heart, is a center line drop as described in the HARNESS chapter.

passes to the outside of the team (either animal) and the cross check runs to the inside of the team as illustrated. (See *Fine Tuning Lines* on page 141.)

Setting up the Lines. With two horses in harness standing side by side, the main continuous left line should pass through the top hame ring and fasten to the left (outside) bit ring. The spliced and adjustable left cross check passes over the animal's back, through the top right side hame ring.

Line Adjustment. (As illustrated, page 173.) With two horses in harness standing side by side, the main left driving line should pass through the ring (on top of the left side hame) and fasten on to the (outside) left bit ring. The left driving line cross check passes

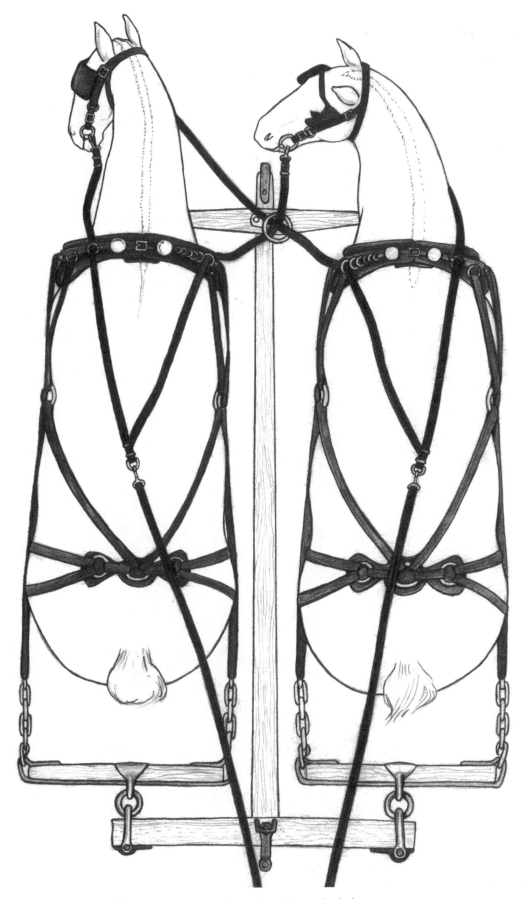

This overhead view shows the position of the team hitched to a tongue.
The lines are passing through the spreaders and a center line drop ring. The harness is market tug style.

over the same animal's back and through the top, right side, hame ring of that left horse. The cross check then passes over to the right horse's left (or inside) bit ring. The right main driving line follows the same pattern in reverse. It passes through the top ring (of the right side hame) of the right side horse and on forward to fasten into the (outside) right bit ring. The right cross check passes over the right horse's back and through the top ring (inside left hame) and on across to fasten into the left horse's (inside) right bit ring.

Button Myers with Patches and Kokomo on walking plow. Photo by Chris Feller

If the team, as it moves forward, walks with their heads pointing in together, check to see if the lines are set up properly with main lines running outside and cross checks inside. If this is correct, adjust the lines by moving the cross check 'forward' at the splice. Now check to see if, when lines are taut and animals moving ahead, their heads are straight. If their heads are facing out, adjust by moving cross checks back a little on the main line. The object is to have the team walking comfortably, straight ahead. If the lines are properly adjusted and one of the animals insists on walking with head in or out, check the shoulder for bruises and the mouth for soreness (inside and out). Some young or green animals will do this out of nervousness and time and work are the only cures.

When skidding a load it may be desirable to have horses work close together, in which case the above described line setup is proper. In other cases, such as working hitched to a wagon, it may be desired to have the animals working father apart. This is accomplished by the use of spreaders (see page 227). Make NO adjustment in the lines. Rather fasten the spreaders to the tops of the inside hames and pass the cross checks through the end spreader rings (instead of the hame rings).

Eveners. For team hitching, doubletrees may vary in width from up to 48" wide down to 30" wide, depending on the job to do. The neck yoke should be the same width as the doubletree.

A logging doubletree with a chain setup.

Photo by Nancy Roberts.

tandem

On the British Isles it is common to see two horses working in line rather than side by side. Such a hitch configuration is called 'tandem.' In situations where the going is narrow and requires additional horse power, this hitch may be needed. It is rarely seen outside of the show ring in North America. Perhaps one reason is that it is one of the most difficult hitches to drive. Normally, if using a wheeled vehicle, the "wheel" horse (or the animal nearest the vehicle) is between shafts and easily accessible, but the 'lead' horse is way ahead and relatively free. The lead animal must be well-schooled, quick to start, a straight mover, quick to stop, and all the while easy to control. It is extremely easy to jackknife this hitch.

The wheel horse is hitched in normal fashion with the leader's tugs hooking into the wheeler's harness, or, better yet, tugs. Ample room must be allowed so that physical contact isn't made between the leader and wheeler, (usually 3½ feet).

Page 232 illustrates the line setup. The teamster will have to handle two lines per hand and keep them all separate and even. This will be done in the same manner as illustrated on page 248 for four head. Notice on page 232 how the leader's lines pass through 'line keepers' which hang from the sides of the wheeler's bridle. These line keepers are the same as

Mic Massey of England logging with a tandem hitch of his Shires. Notice he's driving the leader and the wheeler is following.

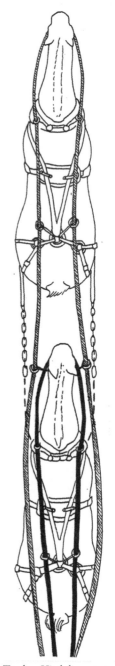

Tandem Hitch line setup.

illustrated on page 225 only smaller. They are normally hung from the bridle, either buckled in or snapped in above the blinders just below the ears.

The driving lines for the tandem consist of two pair of single lines. One is of ordinary length (16' to 20') and the other set is 30 feet or more depending on size of animals, vehicle and/or procedure.

unicorn

This hitch was used in city traffic when a team was not quite enough power. Rather than hitch three abreast, which takes more road space, a third horse was hitched ahead. If the load to be pulled wasn't too big, the lead horse's singletree was hooked to the end of the tongue. If the load was substantial, the lead horse's singletree would be hooked to a cable or chain which would hang parallel to and under the tongue, hitching directly to the same pin as the wheeler's doubletree. In such cases, since an "evener" effect was not had, it was up

The Oregon McInnis Unicorn hitch. Photo by Nancy Roberts.

to the teamster to be certain that all animals were pulling their fair share.

This is a difficult hitch to drive and requires the same exemplary qualities in a lead horse that are required in the tandem hitch.

The lines on the wheelers are set up the same as with a regular team. The leader's lines are set up the same as the tandem leader's, see above.

Unicorn Hitch Lines

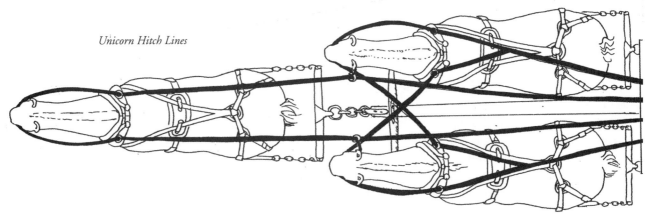

Below: Draft Horse Festival Unicorn Hitches. Photo by Nancy Roberts.

Wilbur Hartranft with Barney, Butch and Billy. Photo by Chris Feller

three abreast

This is a most useful hitch for farm field work but is rarely seen in the woods or on the highway because of the extra width. For plowing, discing, harrowing, even spreading manure, three abreast gives an extra measure of power that keeps horses in harness longer. In other words, the work might be done satisfactorily with a team, but adding another animal in the hitch means less strain on each individual. The result is that the animals are fresher at the end of the day and at the beginning of the next.

Different designs of three horse eveners are illustrated in chapter 9, HITCH GEAR.

There are several different possible setups for lines. Page 235 illustrates a system which requires a set of standard team lines (extra long if possible), plus two 38" long check straps. These straps can be made of line leather with a snap on each end and a conway buckle for adjustment. Or, if unavailable, a regular halter rope can be used, being tied on at the hame end. Individual animals and different circumstances may affect the position on the hames that check straps fasten. Normally they snap into the top hame ring which makes the strap run at the same angle as a line would. The way this system functions is that as you pull one line you pull two horses, the third animal being brought around by the center one. This system is sufficient and works well with trained horses. If you're using an unschooled animal, hitch it in at the center position. That way there is a line on both sides of the mouth and full control.

If you are inexperienced or have several green horses, page 236 might be a better line setup. Here a set of team lines are set up with a second set of cross checks. In this way, there is true line contact with both sides of each horse's mouth.

If you have one horse which likes to move out faster than the others, put him in the middle, as it will be easier to hold him back and even.

Gary Eagle opens a contour furrow with three abreast. Photo by Kristi Gilman

Three abreast. Photo by Marianne Johner.

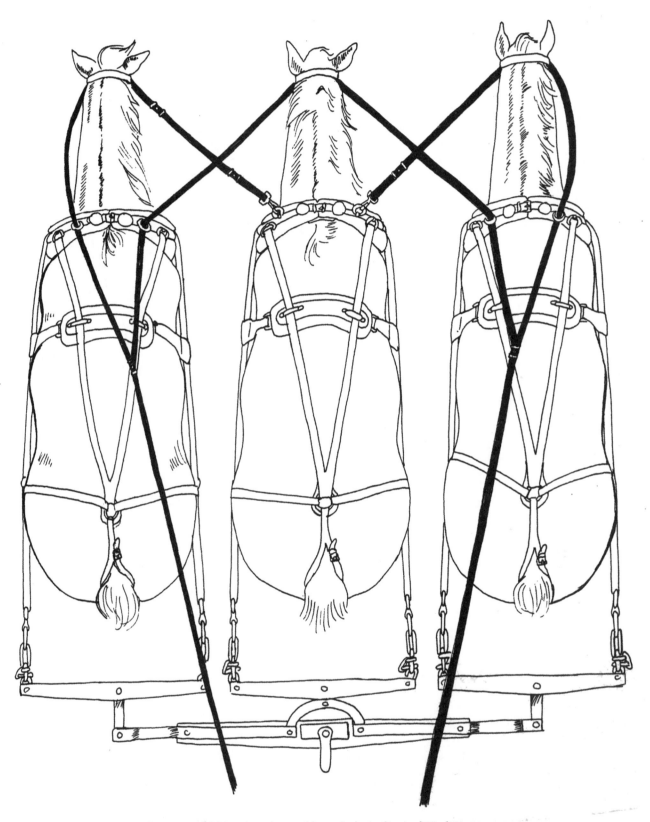

One system of driving three abreast with standard team lines and two short adjustable check lines. These checks are snapped into the bit ring of the outside horse and back up to the top hame ring. Another variation of the same structure has the checks snapped straight across from bit ring to bit ring. This does steer the hitch well but provides no backward restriction to that inside bit ring when stopping.

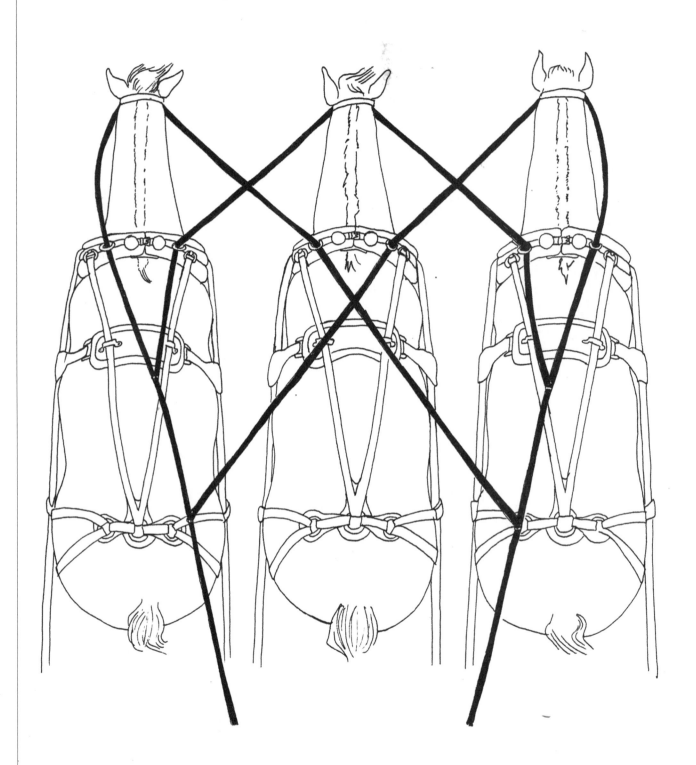

A full line, double cross check, lines systems for three abreast. One design has a short check, 6 foot long, set in the long line at 5 foot 8 inches back. The longer check might be 8 foot long and set back at 7 foot 4 inches. All checks should allow adjustment at the splice of plus or minus 6 inches.

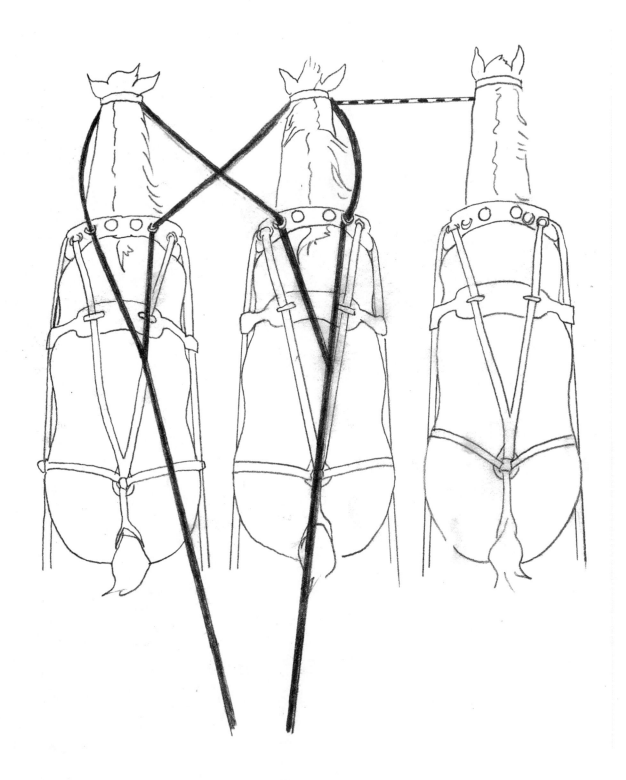

*In Amish communities and parts east, farmers prefer to use
standard team lines with jockey sticks for wide spans. The dotted
line above illustrates the position of the jockey stick which snaps
into the outside bit ring outside horse to the opposing side bit ring
on the horse in the middle. Another path for the jockey stick
would have it go from outside horse at a halter ring (halter worn
under the bridle) to the inside horse's hame.*

A three abreast evener at work on a sulky plow.

Ontario's Aden Freeman driving three Belgians on a 1980 vintage Hochstetler sulky plow. Both photos by M. Johner.

1980 Canadian Work Horse Workshop. Photo by Marianne Johner.

(Below) The Wilder four abreast show hitch of dapple-grey Percherons at the last Draft Horse Festival. Photo by Lynn Miller

four abreast

This is a popular farm field work hitch which will cover a lot of ground. It is important with this outfit to have long lines as when the hitch turns the outside animal will move a long way out and away. Pages 240, 241, 242 illustrate two systems for lines and a subset for jockey stick use. The captions explain those systems.

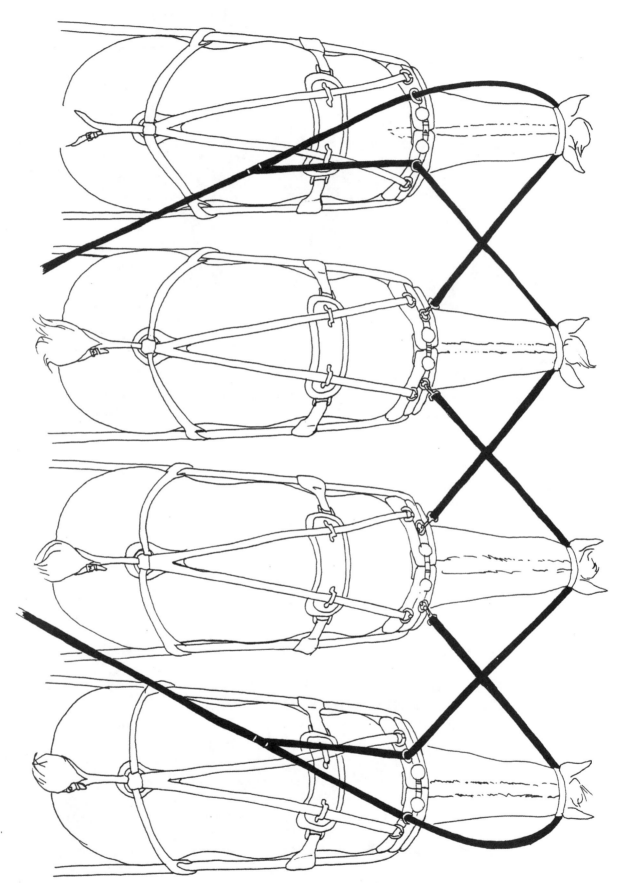

This setup employs a regular set of (long) team lines and four cross check straps. The lines go to the outside bit ring of both outside animals and cross checks off those lines go to the outside bit rings of the inside pair of horses. The cross straps go from open bit rings to top rings on opposite hames. Over time, I have come to prefer having my best horses in the center and youngsters or trouble makers where I can reach them.

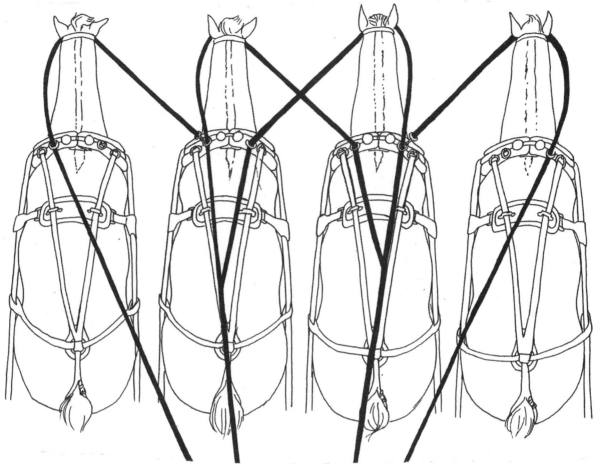

This system requires the teamster to handle four lines. It takes one set of regular team lines, one set of single lines and two cross check straps. The illustration shows how to set the lines up. With this system you have at least one line on each animal, but it does require greater skill to handle the lines.

(Below) Gary Eagle with four abreast in 1977. Photo by MaryLyn Eagle.

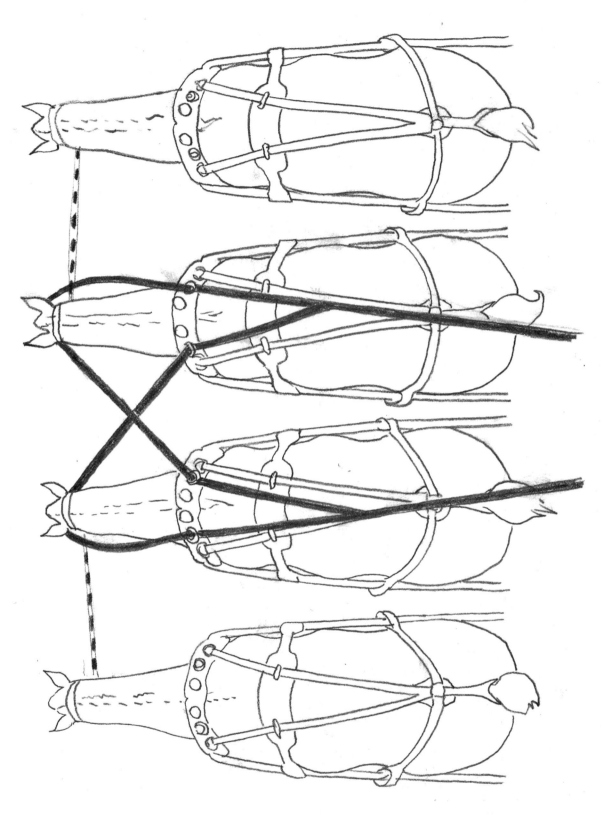

A demonstration of one popular way to rig the lines for four abreast in eastern Amish communities. The outside horses are attached via jockey sticks, snapped into opposite side bit rings (dotted lines). When the middle team turns, the sticks push and pull the outside animals around. I personally prefer not to use this system, especially with young horses which might fight the sticks, because it seems there is a high chance that the horse's mouths will be abused and the work becomes troublesome for them. Also see page 208.

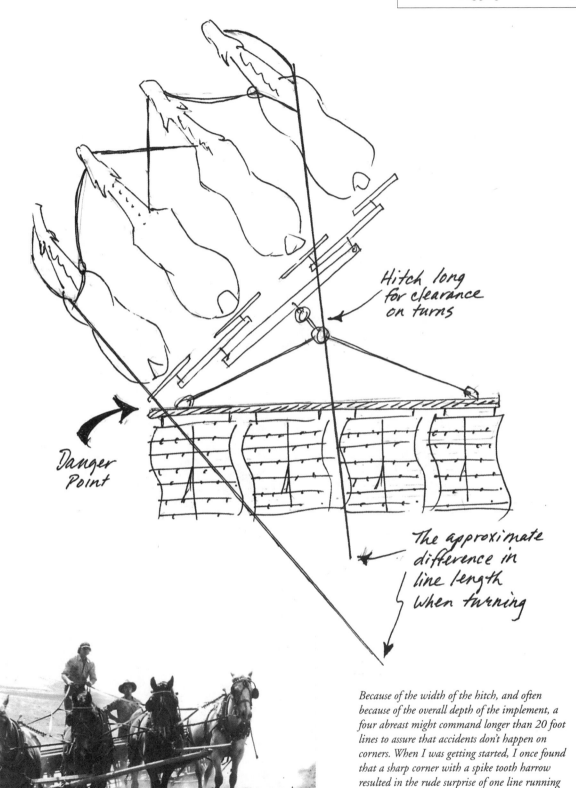

Hitch long
for clearance
on turns

Danger
Point

The approximate
difference in
line length
when turning

Because of the width of the hitch, and often because of the overall depth of the implement, a four abreast might command longer than 20 foot lines to assure that accidents don't happen on corners. When I was getting started, I once found that a sharp corner with a spike tooth harrow resulted in the rude surprise of one line running right out through my fingers.

Gary Eagle and Lynn Miller seeding on the Bear Paw Ranch in 1978.
Photo by MaryLyn Eagle.

The author in 1978 with four up on an Oliver footlift sulky plow.

four up

Working two teams, one ahead of the other, is a popular setup for transport and farm work. Obviously four horses can handle more of a wagon or freight load than a team. And for the experienced teamster, a four-up is an easy hitch to drive. It does require good alert leaders and stout wheelers. Page 245 illustrates a wagon hookup with conventional doubletrees, front and back. In the HITCH GEAR chapter there were illustrations of equalizing four-up eveners which are a must for heavy work. Some of the same driving concerns mentioned in the six-up discussion apply, rather obviously, to this hitch as well. Pages 248 illustrate the principle of driving with multiple lines.

Adie Funk with four Doug Hammill Clydesdales on a small combination harvester.
Photo by Carla Hammill

*Lynn Miller turning the leaders of four up while keeping wheelers straight ahead. This procedure is useful when dealing with narrow headlands up against fences. For in depth information on plowing with horses, please check out the book **"Horsedrawn Plows and plowing"**.*

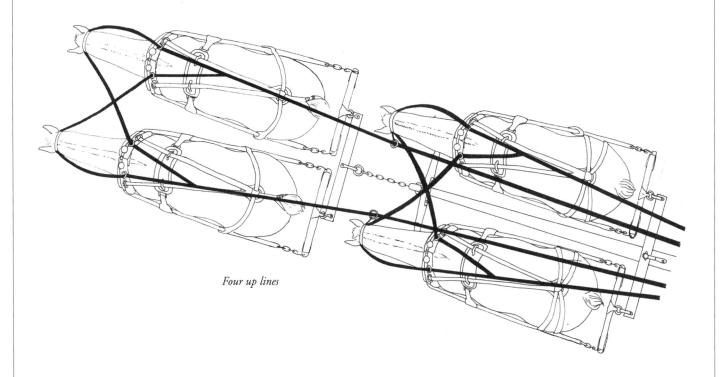

Four up lines

five up

This unusual hitch, employing two horses – set wide – at the wheel and three leaders, is a very effective outfit for farm field work. Because of the position of the wheel horses, the teamster has excellent vantage of the leaders and plenty of room for driving lines. On a large productive horse farm, this hitch will become most popular. The illustration caption will explain the buck-back, tie-in setup for this hitch. This "buck-back" system can be customized to work with any multiple hitch and make it possible to drive with just two lines.

Buck-back. This system works by hooking or tying in all but the lead animals. Buck-back refers to the use of a strap which fastens to both sides of one animal's bit and runs back to snap into the adjoining animal's trace chain (or to the lead chain). In this manner, the "bucked-back" animal is prevented from moving separate of the entire hitch. A tie strap or chain then hooks, preferably to a halter, (rather than a bit), and runs forward to hitch to the trace chain of the animal in front (and to the side) as illustrated several times in these drawings. There are variations of this system as shown on page 257 and, if large multiple hitches see increased use, there doubtless will be many more innovations.

Five up lines showing how one person may drive with just two lines to leaders, and two buckback straps on wheelers and hooked back to lead bar.

Gene Hilty drives a six-up as diagrammed to a trailer-type tractor plow. Gene and Donna Anderson are standing on a mounted platform board on the plow. Note the use of check straps from center horse hames to inside bit of other horses. Photo by Lynn Miller

notes on driving multiple hitches

If you are using the "buck-back" system and driving with just two lines, there are some things to be mindful of with the big hitches. First of all, with the multiple hitches, it is difficult to see where the lead horses are at and too easy to drive into fences or other obstacles (especially if low set). If possible, get as high a seat or platform, as is safe, on your implement so that you can 'see' not only ahead of the animals but also all the rigging! Allow yourself plenty of line length because turns will draw a great deal of outside line. (Of course, be careful to keep lines out of machinery!)

If you are driving big hitches with four, six or eight lines, you will have to have had plenty of driving experience, because you will need the agility that comes from developing instinctive responses to situations. Having to keep more than two lines even, with PERFECT TENSION, is difficult. The basic positions illustrated on page 248 should be a help in keeping lines organized. One of the common mistakes made by beginning big hitch teamsters is 'holding back' lead horses and even swing horses, thereby requiring wheel horses to work harder. Watch that the tugs are all taut and each segment of the hitch is working. This is not a problem with the buck-back system but it can be when driving with full lines. While you want to allow the leaders to pull their share, be careful NEVER to allow the lines to get slack or they may tangle and it is a long way to your lead horses if you are out in the field alone.

If you are driving with four-up or six-up (three teams of two) hitched to a wagon and you are going to back the outfit, care must be taken not to back the lead (and swing) teams into the wheelers.

A five up hitch like the one diagrammed. Photo by Lynn Miller

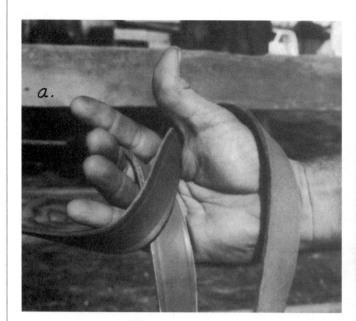

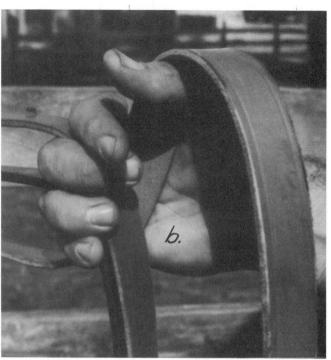

These photos illustrate suggestions of ways to hold, and keep organized, multiple driving lines.

"a" shows a four line system with two lines per hand. In this instance the lead line passes above, the wheel line passes under the hand - up and through the palm - and over the thumb. This keeps the lines simple and untangled.

"b" illustrates six lines (as with six up). Again the lead line is above, the wheel line below (and over the thumb), and the swing line is in between.

"c" illustrates an eight-line system using the same pattern -
    lead, top;
    point, second;
    swing, third;
    and wheel, bottom.

This is a good place to remind you that a buck back system allows you to drive big hitches with just two lines in hand. Unless of course you are using a rope and pulley equalizer which does not employ a "spinal" lead chain or bar to fasten buck backs to.

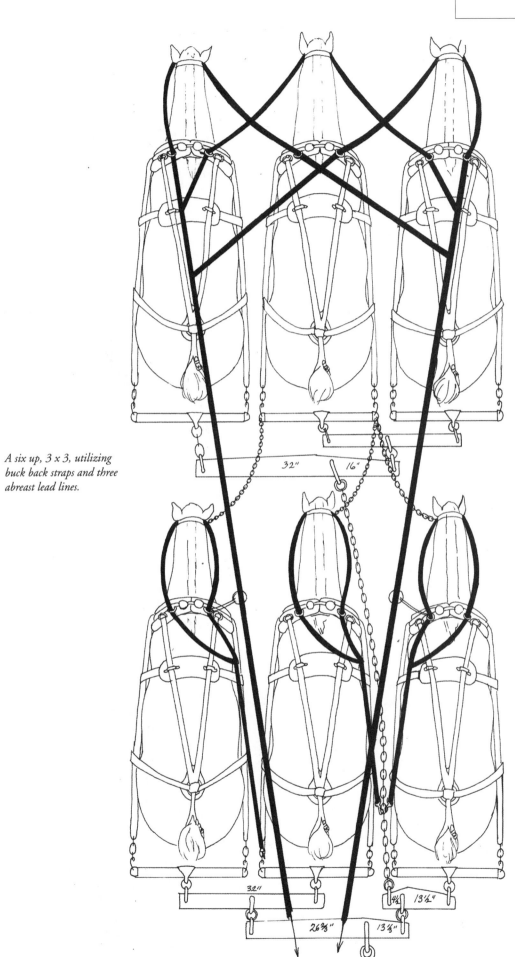

A six up, 3 x 3, utilizing buck back straps and three abreast lead lines.

32"   16"

32"   4½"   13½"

26⅜"   13½"

six up - 2 of 3

This hitch of three and three is, again, a farm field work outfit used for tillage practices. It is an easier hitch to drive than might be readily apparent. In the drawing (see page 249) it might appear that this hitch has a built-in angle on the leader chain that would cause problems. With the offset three horse eveners, as all animals pull ahead, this hitch straightens itself out somewhat. There will still remain a slight side draft depending on the tool being used (i.e., plow or disc. etc.).

The tie-chains and buck-back straps function in the same manner as the five-up on page 246, allowing the teamster to drive with just two lines. This hitch can also be driven with other line systems such as illustrated on pages 235 & 237 or with a system on both the wheelers and leaders.

six up - 3 of 2

This hitch of two, two and two is not a common work hitch because it is more difficult to drive than the six with three and three. But this outfit is the best for heavy highway hauling or wherever a narrow hitch is necessary. The middle team in this hitch is called a "swing" team. This setup can be driven with two, four or six lines. The drawing on the right illustrates six lines. Working with two or four lines will require setting up a buck-back system. If this hitch is to be used on a wagon, it will be preferred to have lines on at least the wheel and lead teams for backing control.

If this hitch is used on an implement or vehicle with a tongue, the wheel doubletree (if an evener) should hitch below the tongue as illustrated on page 245 .

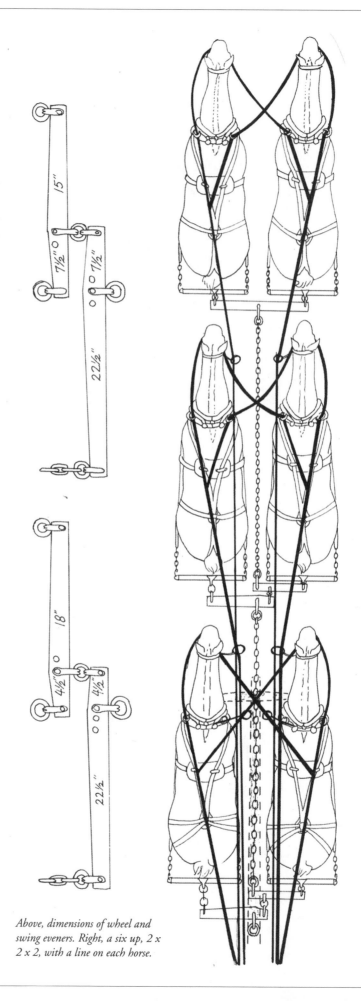

Above, dimensions of wheel and swing eveners. Right, a six up, 2 x 2 x 2, with a line on each horse.

Adie Funk drives Doug Hammill's four-up of Clydesdales hitched by way of a forecart to a combine. Photo by Carla Hammill.

Eight mules hitched (as diagrammed on next page) to a gang plow in Washington state wheat stubble. Photo by Lynn Miller

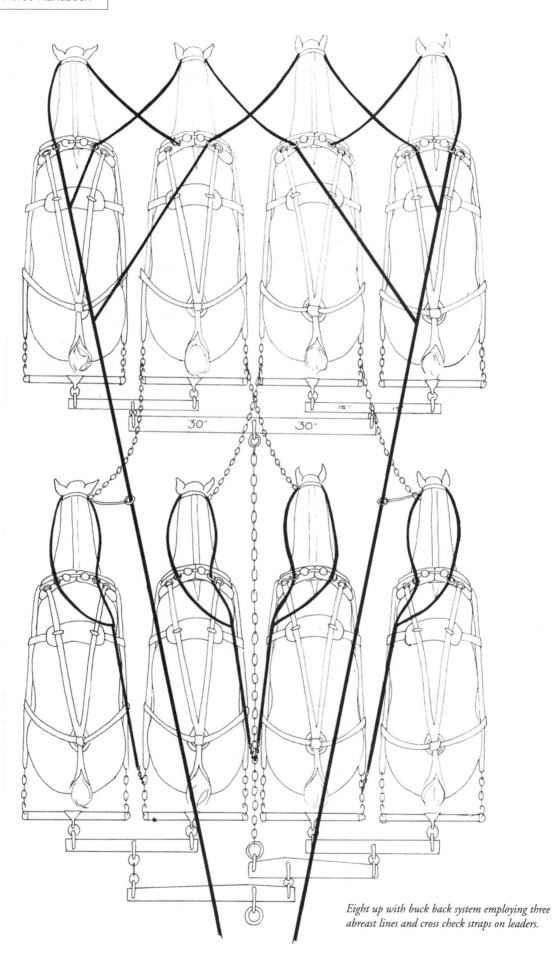

30"   30"

15"   15"

Eight up with buck back system employing three abreast lines and cross check straps on leaders.

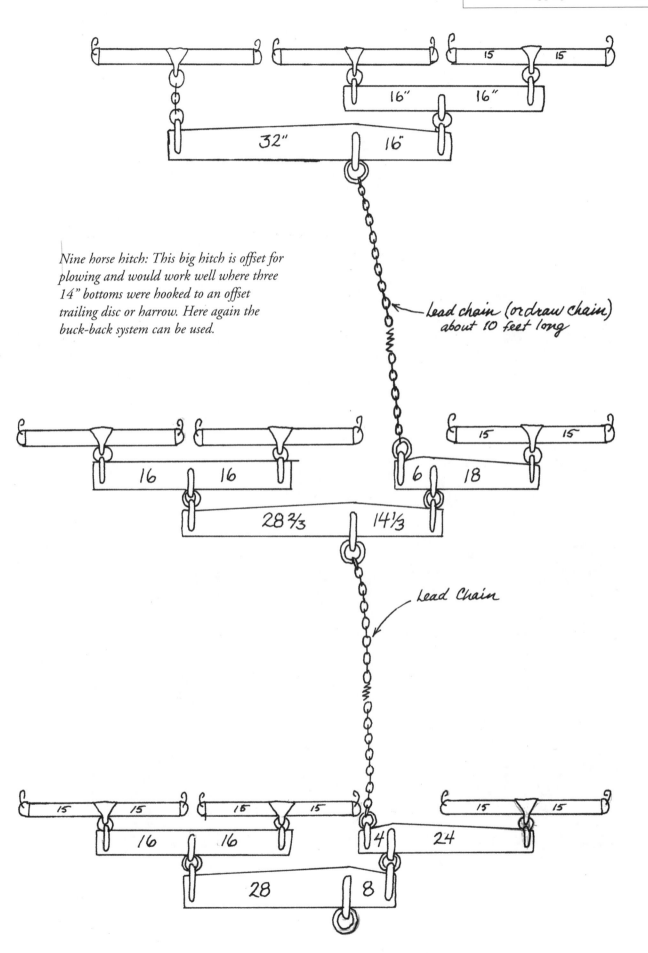

Nine horse hitch: This big hitch is offset for plowing and would work well where three 14" bottoms were hooked to an offset trailing disc or harrow. Here again the buck-back system can be used.

lead chain (or draw chain) about 10 feet long

Lead Chain

In 1979, Don Thomas of Waitsburg, Washington, plowed with this twelve mule hitch. Photo by Lynn Miller

The gang plow used in the above, twelve mule, Thomas hitch.

The lead four of the Thomas twelve mule hitch. Here the check straps run straight across rather than back to the opposing hames. Photo by Lynn Miller

A closeup of the swing team of the Thomas twelve mule hitch. Note the tie chains; one going forward to a lead trace chain, one going over to the bar upright off the evener, and the outside chains running into the neighbor's hame. Also note the line keeper on the far mule's bridles. Photo by Lynn Miller.

All three photos show the twelve mule Thomas hitch as shown in the drawing on page 257.

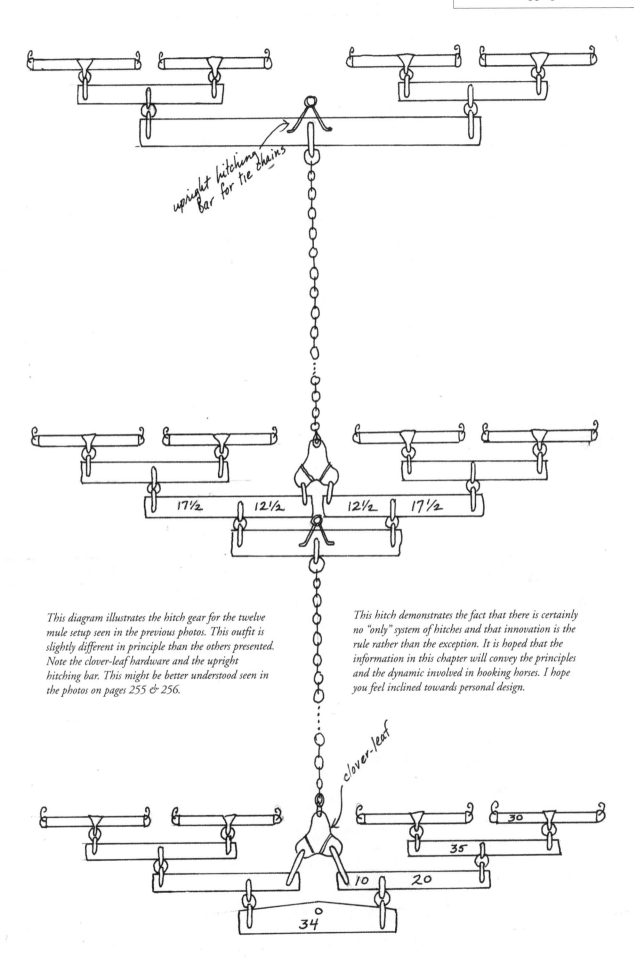

upright hitching bar for tie chains

17½ 12½ 12½ 17½

clover-leaf

30

35

10 20

34

This diagram illustrates the hitch gear for the twelve mule setup seen in the previous photos. This outfit is slightly different in principle than the others presented. Note the clover-leaf hardware and the upright hitching bar. This might be better understood seen in the photos on pages 255 & 256.

This hitch demonstrates the fact that there is certainly no "only" system of hitches and that innovation is the rule rather than the exception. It is hoped that the information in this chapter will convey the principles and the dynamic involved in hooking horses. I hope you feel inclined towards personal design.

I stood on a ladder to get this view of the six Percherons tightening tugs with the White Horse rope and pulley system.

Rope and pulley systems of the last few years seem to be catching on with midwest horsefarmers. Perhaps some day soon an ingenious farmer/engineer will come up with a simple way to incorporate the basic principle of the buck back system and thereby allow the teamster to get by with just two driving lines.

The photo at the below of the Hershberger/Miller twelve Ohio Belgian hitch, shows how a double block is used on the wheelers to equalize the pull of the front eight.

Above, Tom Odegaard, dressed for North Dakota spring, plows with his eight Belgians and a rope and pulley system. Photo by Fuller Sheldon

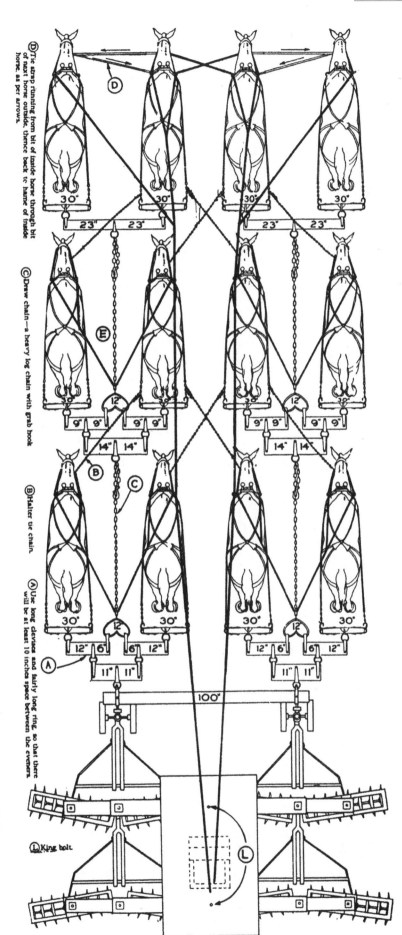

A twelve horse hitch diagramed in our book, Big Teams in Montana.

(D) Tie strap running from bit of inside horse through bit of next horse outside, thence back to hame of inside horse, as per arrows.

(C) Draw chain—a heavy log chain with grab hook

(B) Halter tie chain.

(A) Use long clevises and fairly long ring, so that there will be at least 10 inches space between the eveners

(L) King bolt.

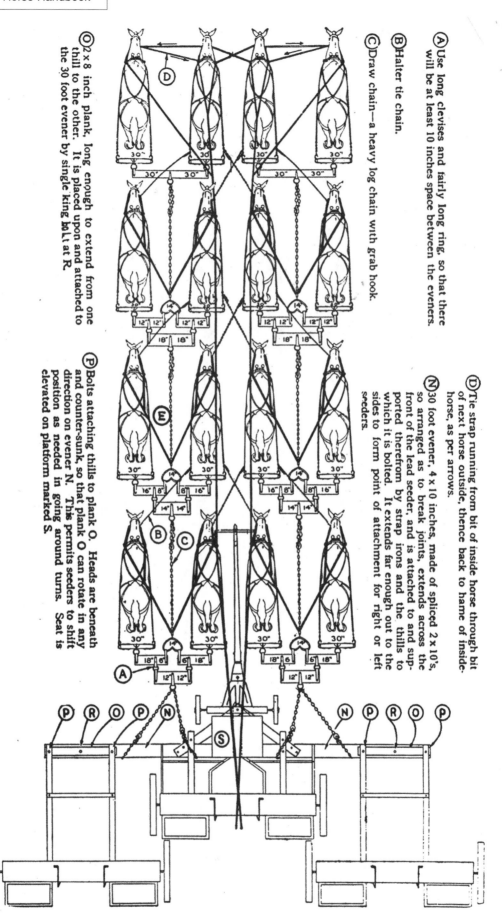

(A) Use long clevises and fairly long ring, so that there will be at least 10 inches space between the eveners.

(B) Halter tie chain.

(C) Draw chain—a heavy log chain with grab hook.

(D) Tie strap running from bit of inside horse through bit of next horse outside, thence back to hame of inside-horse, as per arrows.

(N) 30 foot evener, 4 x 10 inches, made of spliced 2 x 10's, so arranged as to break joints, extends across the front of the lead seeder, and is attached to and supported therefrom by strap irons and the thills to which it is bolted. It extends far enough out to the sides to form point of attachment for right or left seeders.

(O) 2 x 8 inch plank, long enough to extend from one thill to the other. It is placed upon and attached to the 30 foot evener by single king bolt at R.

(P) Bolts attaching thills to plank O. Heads are beneath and counter-sunk, so that plank O can rotate in any direction on evener N. This permits seeders to shift position as needed in going around turns. Seat is elevated on platform marked S.

Sixteen head set up to pull three big seed drills. Buck Back and tie in throughout. Driven with just two lines. From Big Teams in Montana.

CHAPTER TWELVE

HITCHING UP:
Approaches & Procedures

This chapter covers a wide variety of different approaches and procedures for hitching up horses to various implements, vehicles and tools. The possibilities for the application of horse power to work and transport are wide and various. Much the same is true with the varieties of hitching possibilities. We do not pretend to cover every applica-

tion and procedure in this book. The specific technologies and methodologies of logging with horses or farming with horses are deserving of whole books in and of themselves. The information in this chapter is rudimentary and should, coupled with good common sense, put you on your way.

the single horse:

Chapters 7, 9, 10 and 11 have information pertaining to using the single horse covering harness, hitch gear and sundry. If you are to use a single animal

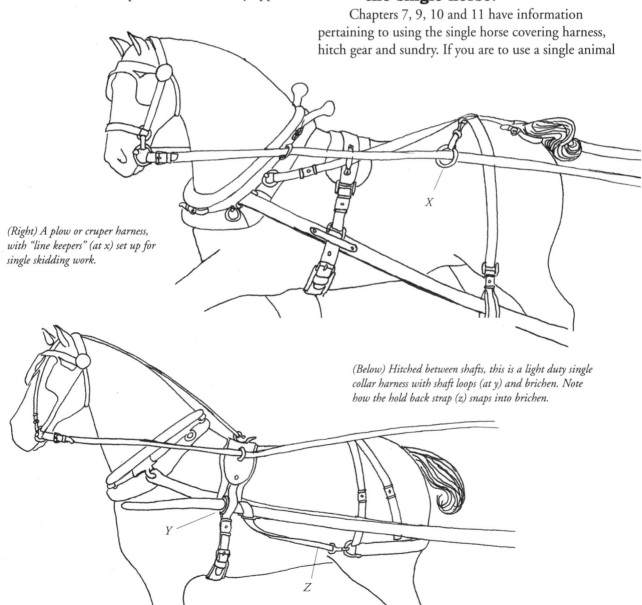

(Right) A plow or cruper harness, with "line keepers" (at x) set up for single skidding work.

(Below) Hitched between shafts, this is a light duty single collar harness with shaft loops (at y) and brichen. Note how the hold back strap (z) snaps into brichen.

That's Ontario horseman extraordinaire, John Male, with a student in 1980. A Belgian gelding is hitched to a light delivery wagon. A standard farm brichen harness has been modified with extra hame straps and quarter straps. The hame straps were used, as a replacement for shaft loops, being wrapped through the harness and around the shafts. The quarter straps were used as hold back straps. Photo by Lynn Miller.

The first few times you hitch horses you should have knowledgeable help, right there with you!

to skid a load (that is to say, to 'drag' something) a plain work harness such as the plow harness on page 261 will suffice. Of course, start with a well-schooled horse, most especially if you yourself are green. If not possible, and both you and the animal are green, have an experienced teamster available to help. There is too much to remember when hitching a horse to a skid load – be it logs, sled, harrow or whatever – you may find yourself confused and unable to give a simple response to a tricky situation. You will need help, on the spot help. It may come down to something as simple as knowing not to ask (or demand) a horse to pull an impossible load.

The best way to gauge what is a comfortable load for a horse is to work up slowly, increase the load gradually. The animal will be able to pull an enormous load (sometimes equal or greater than his weight) for a very short bit. All-day work should be much, much less.

The drag or skid singletree should be equipped with a hook more often than a clevis (see HITCH GEAR, page 187) for easy hookup to chains or rings. The best approach for hitching is to drive alongside of the tool and turn close in front so as to minimize backing up. If you do have to back up the animal to hitch, you may need to pick up the singletree and pull back on it and the lines. Try not to get the animal tangled up and worried. Remember, you are at least ½

of the partnership effort. After hooking, or fastening, to the implement or load, be sure to stand clear of its path if and when the animal pulls ahead. Be especially careful when skidding logs as they may easily roll or slide into you. For this reason it is a good idea to chain two logs together whenever they are small enough to do so.

When hitching the single horse to a cart, wagon or rolling implement with shafts, you can use a harness like that which appears on the bottom of page 261 or you can improvise and convert standard team work harness as illustrated in the previous chapter on page 225.

Hitching to shafts is a little more complex than a skid load. First of all, make sure that the animal will fit the shafts. Have a helper, in the beginning, to hold the horse's head. Then, while the animal is standing, lift the shafts and line them up with the shaft loops of the horse's harness. If the vehicle rolls easily, pull it forward with shafts coming ahead on both sides of the horse. Be mindful not to jab the horse, carefully guiding the shafts into the shaft loops. With each shaft held up by the encircling shaft loop, now, one side at a time, hook the tugs into the singletree and fasten the holdback strap from harness to shaft. The tugs should pass over the holdback straps. While someone is still holding the horse's head, roll the vehicle, carefully, to determine if the holdbacks are set at the proper length to keep the vehicle from running up on the animal. It's a good idea

A team of Aden Freeman's Belgians hitched (by hook) to a flat stone boat. Note how Aden has gathered his lines, keeping them out of the way. Photo by Marianne Johner.

The diagram below shows a sled design which employs a tongue and runners. The tongue allows the team to restrain the load on steep hills and the runners allow some clearance. If wagon tire irons are used on the underside of the runners, rocks or gravel will not wear them down and the sled will start easily. Also, this setup works well, not only in snow, but in wet grass and mud as well. If you want to use this tool in frozen weather, use a pry bar to break loose the runners if they are stuck to the ground. There is a discussion in this chapter on how to hitch to a tongue.

This sled may be made in most any farm shop. It is the first design we came up with. We later developed a plan for a work boat which became a pattern for the Pioneer Work Sled.

Work Sled

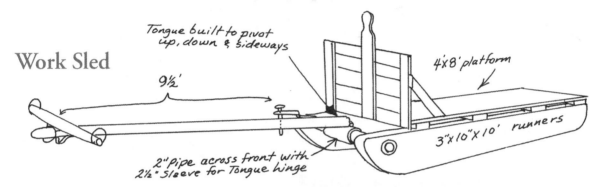

to hook the tugs at the length you wish and adjust the hold backs to take out any back and forth play.

You may have to make harness and hitching adjustments initially, but after that it will be easier and quicker to hitch up. Care should be taken that both the animal and the shafts are strong enough to take any weight that may be transferred, such as in the case of a cart that is front-heavy. The shaft loops should function to keep the shafts from moving much up or down, but there should remain, for the most part, a slight downward pressure.

The Team:

Again, chapters 7, 9, 10 and 11 have information pertaining to the use of the team. That material covers harness, hitch gear and driving. If you are going to use two animals, hitched side-by-side, to drag or skid a load, the harness does not have to include the brichen-

A logging team of Percherons takes a short break. Note the doubletree hooked full out.

In this example, a chain is wrapped around the log to be yarded. And the hook of the doubletree is rigged a particular way, keeping in mind where the log might want to go and where you want the log to go. For example a buried or bound up log can be started easier if the team swings to either side and rolls the log slightly before pulling ahead. And the roll can be caused by hitching over on the opposite side so that the chain pulls the log around. Hopefully you are getting some inkling of just how dangerous this work can be, and why it is that you must always be paying attention, and thinking.

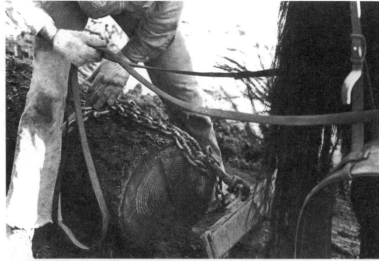

breast strap assembly; a plow (cruper) harness will suffice. However, if all you have is a brichen harness, it will work just fine.

When hooking the team to a skid load of some sort, make sure that your point of hitch is secure, whether it is a clevis or hook. The most common problems come from old cracked or rusty clevises (or clevis pins) breaking when the team is pulling hard or a hook coming undone while horses are relaxing, with the rigging loose. When you hook, it is a good practice to have the team stand just far enough away so that the hitching is done with tugs at least strung out rather than loose. This prevents the horses from stepping over a tug as they move ahead. Be careful while hooking not to be in a position of danger.

Two logs are wrapped together with the chain in such a manner as to tighten the wrap as forward motion is applied. Where possible, it is an excellent idea to put two or more logs together as this restricts some of the rolling action that is so dangerous to the teamster. Both photos by Nancy Roberts.

NEVER STAND INSIDE A HOOKED DOUBLETREE OR EVENER!

Always think about where everything will go when the horses move and where you are in relationship to that trajectory. This is critically important in logging where obstacles and terrain, as well as the hitch itself, can cause unexpected movement of the log(s), maybe right towards the teamster.

Remember, that as the team leans into the collar and pulls, this action can also be translated into a lift. So the shorter (or closer) the horses are hooked, the more likely they will lift the sled, or log, or whatever, as they pull. This can be beneficial in getting a very heavy load started. However, you may be hitched to a tool such as a walking cultivator, where you do not want a 'lifting' action and so the horses must be hooked farther forward.

If you are hooked to an empty doubletree, you will, of course, want to hook it long so that the rigging does not bounce off the walking animal's heels. The

time to shorten the rigging, if desired, is either after hooking or just before as the animals are standing.

At pulling contests you might see a "swamper" (a second person), who carries the short hitched doubletree as the teamster drives. This is done to prevent hitting the horses, because most pullers want to hook to the sled quickly, and because the horses are excited in the competition and safety is good business.

If you are working (yourself green) with a nervous team and do not have an experienced teamster available to help, at the very least have a second able-bodied person there to help hold the horse or correct little problems! Avoiding the accidents from the very start will pay big dividends down the road. The more you work the team (successfully and quietly), the more cooperative they will become.

Walking Plows

The proper adjustment of the hitch is one of the most important factors in the successful operation of horse-drawn plows. In the diagram below "x" marks the center of draft, and "o" marks the point of the hitch. The point of hitch MUST be on a straight

line drawn from center of draft to the point of draft (where traces meet hames) at the collar. If the line is not straight, it will cause the plow to either pull out of the ground or go too deep. If the line is straight, The plow will run easily at the proper depth. the horizontal adjustment of walking plow hitches is relatively easy, provided the doubletree is about the right length and the

plow share is properly set (with just the right amount of suction under the point). If a right-hand plow is not taking a wide enough land, the clevis may be moved one or two inches to the right. If taking too much land, an adjustment of one or two notches to the left will bring the desired results. A left hand plow is adjusted in exactly the opposite manner.

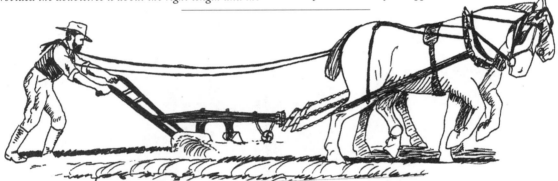

If you are plowing for the first time and without help, it will seem an impossible procedure to master. Experienced horses, a helping teamster and a clean furrow would be ideal circumstances for the beginner. If this is not possible, try at least to have some of the first furrows started for you. If that is not possible, have someone either drive or lead the team until you have a good clean furrow for the horses to follow.

When plowing with a team, the opening or crown furrow is the most difficult. It is best to have some understanding born of experience before you start. One horse will be walking in the furrow and one will walk on the land. For this reason, it MAY be necessary to make a slight adjustment in the lines to compensate for one horse being taller. We recommend the book **Horsedrawn Plows & plowing** by L.R. Miller.

The lines are best tied together and put over one shoulder and under the opposite arm. If an accident of some sort should occur, the teamster, by tucking his head, can easily get away if one line is over and one under. The length of lines depends on the entire hitch and the length of the plow.

To start plowing, lift up on the handles and have the team step ahead. The plow will suck down in to the ground and find its proper depth without your help. As you move ahead, slight pressure on

either handle will cause the plow to move to the opposite side. Increased pressure to either side will cause the plow to come out of the ground. Pushing down on both handles will cause the plow to come out. Pulling up on the handles will cause the plow to go down deeper.

The rule with the walking plow is RELAX! Let the horses pull it and go with the plow, making minor adjustments. If you're pushing, as in the drawing in the middle, you are causing yourself grief. You should be comfortable as the drawing above portrays.

Riding Plows

The diagram above illustrates the correct up and down adjustment for riding plows. As in walking plows, a straight line from "x" (point of draft or center of load) through "o" (point of hitch), to the point of draft on the collar is necessary. When hitching horses strung out, the hitch at "o" must be lower than when using animals abreast. The teamster should try to get the horizontal hitch point as near as possible in direct line between the center of draft and the center of power, when the plow is running straight and horses are pulling straight ahead. To find center of draft, measure total cut of plow. Half of the total cut is center of the cut. Measure to left of center of cut 1/4 to the width of cut of one bottom to get center of draft.

Comfort to horses when plowing depends on the adjustment of horses and harness. For best control, horses should work close together. The hitch being in the right holes on the plow, long tugs give horses more room and tend to make the plow run easier and steadier. If trace carriers, or mud straps, are used, they should be adjusted so that they hang free. If these carriers pull up on the traces, they will change the angle of draft causing the plow to run unsteady and they'll put extra weight on the horses' backs. Ray Drongesen plows with three abreast on a frameless sulky plow.

There are hundreds of different models of dozens of different makes of the old riding plows. Add the dozens of new plow models and you may see why I chose to author an entire volume on *Horsedrawn Plows and plowing*. For indepth information, we encourge you to look to that volume.

For basic hitching purposes, there are a couple of important variations. There are riding plows with steering tongues and some with no tongues. Also, there are frameless plows and frame plows. The single bottom riding plow is called a "sulky." The riding plow with two or more bottoms is called a gang plow unless there are two bottoms facing opposite directions in which case it is a two-way plow.

When all parts of a riding plow are properly adjusted, the combined weight of the plow, teamster and the furrow being turned is carried on the wheels. Shares must be sharp and properly set. The entire bottom must be well polished for good scouring. Rolling coulters or jointers, if used, must be sharp and in good condition.

Pictures of plows appear in Chater Sixteen.

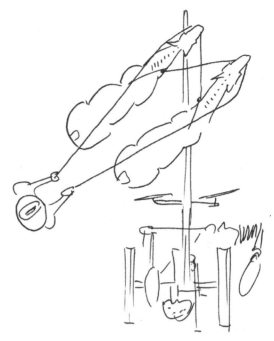

Drive the team over the tongue swinging the pair as the right horse steps over. Try to position the team so that a minimum of backing is necessary to get to hitching position.

hitching to a tongue

Wagons, mowers, corn planters and all sorts of wheel vehicles and implements are often equipped with tongues that function in conjunction with brichen harness, as a restraining, backing and/or braking system. There are some procedural considerations in hitching a team or three abreast (or even four) to a tongue which must be followed if safety is a concern.

(Here, again, we heartily recommend that you educate yourself fully before attempting these procedures. Find a knowledgeable teamster to help you the first few times. And get copies of *Training Workhorses/ Training Teamsters, Horsedrawn Plows and plowing, Horsedrawn Tillage Tools, Haying with Horses.*)

TEAMS: First of all, check to be sure your equipment goes together, is in solid shape, and that the animals are going to fit the equipment. Next, put the doubletree on the implement and set the neck yoke alongside the front end of the tongue. If the tongue lays on a comfortable angle from vehicle to the ground, the team can be driven over to hitch. If the tongue is 'stiff' (hanging in the air), or the angle originating too high, it may be necessary to either back the team or lead the animals one at a time to their positions. Let's take each of these considerations one at a time.

With an experienced team (and a helper), ground-drive the team slowly up to the tongue and ask one of the animals to step over the tongue while you turn them close in. The result should be the team in proper position for hitching with little or no backing necessary. It may take a lot of practice even with a good team. Remember that when you are ground-driving the team, there is no physical restraint preventing them from spreading apart at the rear when backing. Make sure the team is standing straight before asking them to back; that will help some. There are those who actually tie the two together from brichen to brichen to prevent this 'spread.'

CAUTIONARY NOTE: The unintiated teamster may back willing horses into a calamity. As the horses back up it is only natural that they should look towards each other, with no restraint at the rear, they tend to fan. Two things might happen at this point. First, they could literally turn themselves inside out and render the team lines useless. And second, in the process of fanning apart, they may pull a line splice into the hame ring and hang it up there, making it impossible to turn or stop the horse. Back up unhitched horses no more than two steps at a time.

With the team standing in the right place **ALWAYS HOOK UP THE FRONT END FIRST!** (Your neck yoke should be slid over the end of the tongue and secured there with a small clevis, piece of wire or at least a string so that it does not accidently fall off as you are hitching.) Fasten the neck yoke to the breast strap-pole strap assembly. as illustrated in various places throughout this text. After the front end is hitched, hook the tugs. Hook them slightly taut. It is a good idea to hook both inside tugs first, then the outside tug on your side. If you are alone, be very

careful moving around in between the horses and the equipment. It is always safer to carry the driving lines, and walk around the implement to hook the last tug.

If the wagon has a 'stiff' tongue, it may be necessary to back the team, carefully, with the tongue coming in between; or to simply lead each animal to position, then hook up the lines, and then the neck yoke, and finally the tugs.

CAUTIONS:

Always hook the neck yoke first, NEVER hook the tugs first!

Always have lines hooked up and in hand or easily reached when hitching!

Never step inside and/or between the eveners and the horses!

Never trust anyone to hitch or hook an out-of-view part or piece without checking it yourself!

Have lines in hand when mounting implement or vehicle!

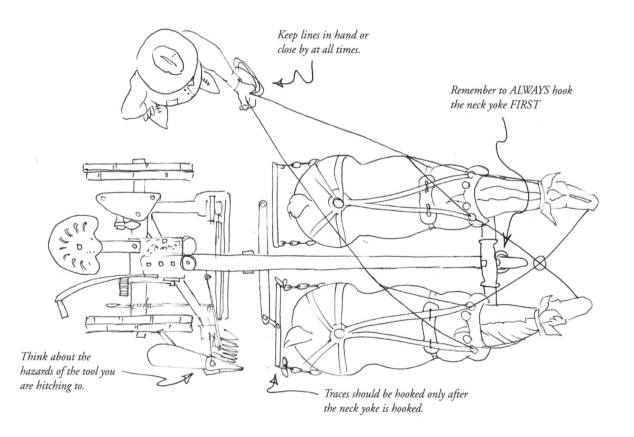

Keep lines in hand or close by at all times.

Remember to ALWAYS hook the neck yoke FIRST

Think about the hazards of the tool you are hitching to.

Traces should be hooked only after the neck yoke is hooked.

The legendary Herman Daniel with three Belgian mares. Photo by Nancy Roberts.

three abreast:

When hitching three to a tongue, obviously two animals will be to one side and one to the other. For this reason, it makes sense to drive the single animal over the tongue rather than two horses over. If the three are to be hitched to a regular center-line tongue, a special offset evener will have to be used. See pages 200-207.

This evener (as illustrated in HITCH GEAR) is built to hitch in line with the tongue (1/3 over) and still have all animals pull even. The design employs offset lever support to get the job done.

When hitching three abreast to a riding plow such an evener isn't necessary as the tongue is adjustable and separate of the hitch which is also adjustable. So a regular three horse center fire evener can be used. (See page 191). The same hitching principles apply. NECK YOKE FIRST, then tugs.

four abreast and big hitches:

Four abreast hook the same way with tongue running in between. Principles and rules are the same. It is, of course, more difficult to back three and four abreast than a team.

(Right) In 1976, the author unhooks a double tree from a disc harrow. Notice that the lines are kept in hand. Be mindful of the path of the implement and where you stand should the horses be startled and step ahead.

When hitching multiples with wheelers (and possibly a swing span) to a tongue, hook the wheel team completely before hitching additional animals ahead. The first thing to do is have lines set up and in hand. Then hitch the front end of the wheel team, then the tugs. Do the next team ahead and so forth. Have help until you've got your animals 'worked down' and a system set up for each hitching.

(Right) Pete Weimer with six Belgians in Montana 1980. Photo by Christene George

(Below) Lynn Miller demonstrating putting together six up plow hitch at North Dakota draft horse meeting. The Clydesdales belonged to the Tweeten Brothers.

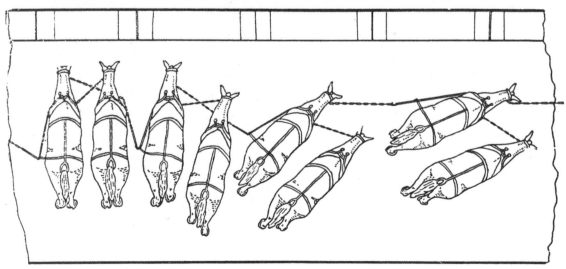

Drawing illustrates one way to go to the field with multiple hitch horses. In this case, teamster leads lead team and each subsequent span, or team is tied to tug end and must follow. Another way is to drive leaders and lead remainders from behind.

rope and pulley equalizer

Today's incarnation of the rope and pulley system is perhaps best shown through the rigging designed and sold by White Horse Machine of Gap, PA.

Below left, the ropes are being laid out and hooked to the wheel evener, here a four abreast, in anticipation of the eight horses, 4 + 4, being lead or driven over and hooked. The wheelers will have the short side of the continuous rope hooked directly into the trace chains, while the leaders will have the longer section hooked into their trace ends. Carrier chains will drop from the breast strap ring (as in photo below) to carry the ropes.

Below left: In this photo you see the pulleys from the twelve horse swing team hooking into the ends of the double block used on the wheelers.

Below: A closeup of the wheel eveners where you can see the rope hooks fastened to the end of the traces. Then you see the rope passing around the pulleys and forward to carrier chains and on to the traces of the horses ahead.

*A split apart fish eye view of the twelve horse wheelers
with rope and pulley equalizers.*

*One of my concerns with this system are the hazards inherent to a web of ropes moving forward
through the entire hitch and not allowing safe access for correction. In this picture you see that a
horse has stepped between the rope. Should that horse step forward the rope would burn his lower
leg and perhaps cause some anxiety. In order to fix the problem, the horses must be held and
someone has to crawl in and deal with the mess.*

On the right: This is a view of the apparatus for the twelve horse hitch at the heads of the wheelers. Notice the double block going back to the evener. And note the single rope going forward, along side the swing team and on to the leaders.

The carrier chain visible on the near wheel horse should be slack like this when the pull is under way. If this chain were tight it would cause the ropes to pull down on the collar.

The harness is a synthetic sidebacker style which explains the loose extra strapping along the side of the animals.

Notice that the lines are running to center teams and that the outside horses are fastened in with jockey sticks.

A clear view of the wheelers stepping into the rope and pulley equalizer.

And a view of what it looks like standing on the four bottom plow looking out over the Hershberger/Miller twelve Belgian hitch at the 2003 Horse Progress Days held in Mt. Hope, Ohio. All photos by Lynn Miller.

Joe Van Dyke in 1976 at the Dufur Threshing Bee parade in central Oregon 1976. Notice that a chain is used to move evener forward so that it doesn't lift up on harrow. Photo by Lynn Miller.

Gary Eagle in 1977 at Chesaw, Washington with five abreast pulling disc and spike tooth harrows. Photo by MaryLyn Eagle.

CHAPTER 13
COST OF USING HORSES

Lynn Miller and Ray Drongesen in 1977. Photo by Christene George

From 1867 to about 1930 the USDA conducted studies and surveys on the animal power question and published numerous excellent, although often slanted, bulletins offering the findings to the general public. This writing draws on some of the basic statistical findings (updated) of that body of work with statistics from current practices and information from personal experience.

This writing is meant to answer questions about the cost of using horses. It also answers some questions along the way about the number of horses needed for certain acreage and vice versa. This chapter goes into some detail but it is centered around certain common practices, crops and economies and the reader may have to work out dollar and crop translations modifying the information for his or her particular circumstance and area.

In computing the actual gross costs of using horses for motive power on the farm it is important to account for many items which are not direct cash outlays. For instance, feed grown on the farm or labor performed by the farm family may not require an actual cash expenditure but they are true costs and need to be computed in. The actual dollar costs attributed to these sorts of entries are going to vary remarkably

depending on the attitude of the farmer and the total circumstance. Meanwhile, persons needing to make practical decisions about the future of work horses on their farms are going to be most particularly concerned with actual cash outlay required. For this reason we will include this as a separate category.

The use of horses has certain offsets if the farmer chooses to take advantage of them. For instance, manure and the raising of foals for replacement or sale have definite values which serve to offset either the gross costs or the cash outlay. For this reason, we also include their effect by offering categories of net cost (gross costs less offsets) and net outlay (cash outlay less offsets).

There are several intangible values or disadvantages, relative to specific operations, which either cannot be computed or which we choose for the sake of simplicity to exclude. We will mention those at the close.

We must assume that the use of work horses presupposes a deliberate choice of a mixed farming system of moderate size as versus a large monocultural practice. This is not to say that horses could not supply dependable power for a large single crop operation, but rather that the true economy and practicality of the horse is best realized on a farm with various crops in rotation including crops suitable for horse feed.

Understanding and controlling the cost of maintaining work horses with the concern on the one hand of realizing maximum work efficiency while on the other hand keeping costs to a minimum will result in a greater share of production income returning to the farmer. The fact that much of the expense of maintaining work horses is made up of feed produced on the farm makes it easy for the farmer to be less concerned about cost. But, insofar as this has a direct relationship to the amount of commodities available for sale or to feed to other stock, it has a direct bearing on the total farm profit.

By USDA census, on January 1, 1921, there were 24,663,000 horses and mules on American farms. These animals had a total value of $2,255,991.00 and were maintained primarily for the purpose of furnishing farm motive power. The American Horse Council reported in the spring of 1979 that there were 8,200,000 horses in the United States. These animals have an unknown value and are used primarily for sport. The regenerative capacity of horses is tremendous and, because of various interests, total numbers have tripled in 20 years. Today, 2003, there are more horses than were counted in 1921.

The gross cost of keeping work horses is made up of feed and bedding, labor, interest, stabling, use of harness, shoeing, depreciation, and miscellaneous charges. If we subtract the offsets from this gross cost the resulting net cost would be the cost of the work performed.

From 1914 to 1918 the annual cost of keeping work horses on corn-belt farms of approximately 160 acres usually amounted to $450 to $750 per year. The average per horse per year cost being $99.21. So, obviously, the farm which could efficiently complete its work with four head of horses (160 acres) had a lower overhead (and higher return) than a farm which used seven head of horses.

A farmer/stockman today might have six mares and a stallion (and growing foals) and perform the field work for less money than another farmer with four geldings. For the farm with available space, top consideration should be given to working mares and raising foals for profit and replacement. (And it is not necessary that the farmer raise registered purebred stock, as the market for good grade work horses is excellent.) It is important to understand the nature of costs involved in using work horses so that they might be efficiently controlled. Some of the costs are of minor importance, others of great importance.

The table on the bottom of page 281 shows the categories of cost and the percentages of the total. These figures are updated from a 1921 USDA survey and are based on a 160-acre mixed crop and livestock farm using six geldings.

For the sake of comparison, the table on page 285 shows projected cash outlay on the same farm scale but with a work unit of six brood mares and one stallion.

Feed

Feed and bedding combined (see tables on pages 286 & 287) is the most costly category at 39% of the total. Yet when comparing the tables on pages 285-287, feed becomes an area of expense with which the farmer might have tremendous flexibility. For these two reasons, feed as a category deserves first consideration in attempts to control or reduce costs.

On page 285 there are notes showing how cost calculations were made. If you are already using horses, you may find fault in these calculations. Some differences can be traced to substantial variations in climate, available feed crops and farm economies. It is, however, more likely that a misunderstanding of the horse's digestive system and the best principles of feeding for work are the root cause of many farmers overfeeding (or, in effect, wastefully feeding) work horses. See FEEDING chapter.

The best methods of feeding work stock should always have first consideration on a well-managed farm. The quantities of grain and hay must depend on the kind and regularity of work, the speed at which the work is performed, quality of the feed, age and condition of the horse, and the keeping qualities of the individual animals. Although the exact quantity is variable, a good, practical guide for the farmer to follow in feeding the horses is to allow 1.1 lbs. of grain and 1 1/4 lbs. of hay per 100 lbs. of live weight when the animal is performing moderate work. For horses at hard work, the grain should be increased to about 1 1/4 lbs. daily per 100 lbs. live weight, but the hay should not exceed 1 1/4 lbs. daily per 100 lbs. live weight (unless the hay is of very poor quality).

Quite contrary to the boasts of many old-timers and not-to-be-outdone newcomers, who have the average weight of any respectable work horse at at least a ton, surveys conducted 40 and 50 years ago and personal observation over these last thirty years indicate that the average weight of most of the horses which truly work is about 1400 lbs. The range is anywhere from 500 lbs. (nod to our pony friends), to 2500 lbs. per horse. Far and away the majority weigh in at 1200 to 1600 lbs. The table on page 285 shows the feed requirements of 1300 to 1400 pound work horses. Our horses may not be as big as we like to think they are. That's good news in our cost analysis because it means we don't need to feed as much.

The amount of feed required by horses varies, not only from season to season, but also from day to day. Just because four horses are being used at heavy work does not mean that all the horses should receive a similar heavy grain ration. The observant teamster will find that a certain animal may perform well and hold his weight on slightly less grain. It is the little daily reduction of grain that counts up for a large saving during the year. Also making individual determinations of need will assist in keeping the horse in working condition.

As the spring work season opens up, the amount of grain must be gradually increased so that the horses are ready to receive a full grain ration when the heavy work begins.

Aside from the idle winter season, there are times during the summer when at least some of the horses are not working. During such periods, the use of good pasture in place of the grain and hay ration is not only an economical practice, but also will have a good effect on the digestive system of the horse. If the pasture is short or poor, it may be necessary to supplement to keep the animals in good flesh.

If a horse is on night pasture (an excellent practice) and worked during the day, the regular grain should be fed, depending on the nature of the work. If the pasture is good, the horse will consume only a small amount of hay and if a large amount is kept in the feed manger, a considerable portion will be wasted under foot. At all times, the horse should be fed so that he will utilize all the feed put before him, which will aid directly in reducing costs by preventing waste.

Many farmers are wasteful in the feeding of hay. It is common for farmers to keep the manger filled with hay at all times. This is not only a serious waste but experts say it is also detrimental to the horse's health. The horse has a small stomach and a touchy digestive system. He cannot take care of great quantities of roughage during the working season. If you feed too much hay to your horse prior to work, it will have a bad effect on the respiratory and digestive systems, and is the cause of excessive sweating and fatigue. Only a small amount of hay should be fed to horses in the morning and noon meals; the greatest amount should come at night. It is also an excellent practice, if possible, to allow the animal to eat some small quantity of hay before placing the grain ration in front of him. The roughage in this instance acts to slow the digestion of the following grain, allowing the horse to derive more nutrients.

The use of farm raised feeds which aren't readily saleable is to be recommended. If farming practices provide a quantity of good straw, dried corn stalks or stalk pasture, their use as horse feed will help to reduce the feed expense and permit the sale of hay or its use to feed other stock. The practice of letting the horses have the run of straw stacks (if you are lucky enough to have them from threshing) or stalk fields during the fall and winter not only results in saving of more valuable feeds but tends also to make the animals more hardy and in better shape for spring work.

To obtain the best results in feeding, it is imperative that rations be properly balanced and supply adequate protein for the building of tissue and the supplying of energy. The less protein in a given ration, the more feed required. A little time spent in the calculation of rations enables the feeder to provide the proper nutrients which benefits the horse and can often reduce the feed bill.

Chores

The USDA 1921 study on the cost of using horses surveyed 279 farms and found the average amount of time spent feeding and caring for seven head of horses per farm amounted to 467 hours per farm (or 66 hours per year per head). On some farms the amount went as high as 125 hours per year per horse.

Chores are nearly always done either by the farmer himself or by members of his family without actual cash outlay. Some reduction in time spent can be had by having convenient feeding and stabling arrangements so that handling horses and moving feed can be done in as short a time as possible.

Depreciation

The net decrease, or depreciation, in the inventory values of work stock on the farm is a calculation of the per year value of the total cost of replacing worn out stock. On the farm which depends exclusively upon gelded horses for work, there is no escaping this cost item even though it may not represent an actual cash outlay until the day of need.

Photo by Nancy Roberts.

One system used by some farmers to greatly reduce the true depreciation is the purchase of young, three year old, stock which is worked until five to seven years old and then sold (in the prime of life and possibly with excellent training). The proceeds of the sale are then used to purchase young stock again.

The item of depreciation can be (and is) lessened or removed altogether on some farms by the raising of foals. Foals increase rapidly in value and continue to increase after trained for work, until the highest value is reached at about seven years. Thus there is little or no depreciation other than that caused by disease or injury until age seven or eight. With today's horse meat prices and draft horse prices, farms which use brood mares for work and to raise foals for replacements and sale are showing steady appreciation in the total value of work stock rather than depreciation.

Strong young fertile mares can be worked steadily from one week to one day before foaling and, if no complications arise, can return to work within a week. Care should be taken that pregnant mares are not required to pull extremely heavy loads under slippery conditions as a fall or wrenching slip might cause problems. Most vets, however, will tell you that a mare which is worked steadily will have a better overall physical condition which makes birthing easier. After foaling, the mare's diet must be balanced and adequate to produce both work and milk without loss of body tissue.

The best methods of feeding and care of work horses, keeping them in good condition, is an important factor in reducing depreciation charges. Overheating or overworking the horses and the lack of proper attention often makes horses unfit for hard work when they are needed most. Permanent injury or chronic disease may be the result, requiring premature replacement.

Stabling

Stabling cost concerns only that part of the building in which horses are housed and feed is stored. On farms having low-priced or solid older structures, the cost of sheltering is a small item; on other farms, where considerations are urgent and expensive, it is quite another thing.

A small part of calculations included in the tables are actual cash outlay.

Harness Costs

The maintenance cost of harness becomes properly a part of the cost of horse power. It is an item of small importance and can be reduced some with good habits of use, care and storage.

Shoeing

Shoeing and, to some degree, hoof trimming, are direct cash expenses on most farms. The cost of hiring this work done is increasing as farriers are justifiably insisting on a fair wage. Costs can be kept to a minimum by the farmer learning to trim his own horses' feet and by shoeing only those horses which require corrective or preventive work and those which must go distances on gravel or asphalt roads. Horses with strong healthy hooves will be best off if left barefoot to do field work.

Miscellaneous

This item is composed of veterinary services, medicines and salt. It represents a direct cash outlay and the amount will be affected considerably by the time and service required of the vet. With good preventive care and a well-educated farmer (in vet matters) this cost can be held down.

Manure Offset

In computing the manure credit, I used the

average actual nitrogen produced by one 1400 pound horse over a year at 160 lbs. (the total manure tonnage came to 7.5 per year). At a chemical market value for actual nitrogen of $27.50 per hundred weight, the value of manure per year is about $44. Since it is difficult to sell manure, the value used does not represent a cash credit but rather a cost offset. An offset of this nature will vary in amount according to the individual manure handling practices and the relative value to the individual farmer of the fertilizer.

Aden Freeman and friend. Photo by Kristi Gilman-Miller

Cash Costs

Only part of the items discussed here can be considered as being either cash expenses or involving materials having a sale value. The hay, straw and grain consumed may all be salable. Stover and pasture could be considered byproducts of the farming system and may be difficult to sell. A percentage of stabling and harness costs could be classified as cash expenses. Hoof care and miscellaneous vet expenses are most likely cash expenditures. If we total the feed, a percentage of stabling and harness ($10) and all of hoof care and vet expenses, we come to a cash cost of $401.82 per year per horse. If we adjust the total to reflect actual cash out of pocket and assume that all feed is grown on the farm, we come to a total of $110 per horse.

Cost of Horsepower

The daily or hourly cost of horsepower on any farm is dependent upon the number of work horses, the cost of keeping them and the number of days or hours worked in the year. As we've shown, the cost of keeping horses can vary dramatically. In the 1921 survey, 279 farmers averaged 723 work hours per horse. The more horses per farm (of the same size) the fewer the hours per horse and vice versa.

Using an average of 723 hours of work per horse the average cost per hour would be $1.03 per hour (or acre). This is, of course, the high end of figures presented in this chapter. For a suggestion of the effect of the variables in cost, look to the tables which follow.

Obviously, if all things were equal, the horse that worked 1,000 hours would be less costly than the horse that worked only 500 hours per year. Many types of American farming are such that some of the horses

maintained throughout the year are needed only to perform necessary work during the crop season or at rush periods so that the amount of work done by a group of horses is not limited so much by their yearly capacity as by the distribution and the amount of work to be done.

The best combination of crops and procedures for a given farm may require a large number of horses for only a short period of the year, even though many of the horses are idle during the greater part of the time. The average hours worked per day or per year might be very low and still be a justifiable result of good management for that particular farm.

There are many exceptions to the cost rules we have employed here and they are only meant to be guidelines to help farmers make practical management decisions. If nothing else, the spread in possible cost numbers indicates the extent to which a careful farmer can affect his costs.

Referring again to the 1921 study, the average number of acres of crops tended per horse in different sections of the corn belt varied from 18 to 24 acres. Obviously, the higher the number of crop-acres per horse the lower the cost per crop-acres of horse work performed. But that is a dangerously simplistic mathematical distinction and serves only to indicate horses needed per acre. It should not be used as a measure of efficiency as there are far too many exceptions in this regard. One horse for every 18 to 24 acres of cropland farmed is an excellent rule of thumb for people desiring to compute future need. On a mixed crop and livestock farm with woodlot, ponds, swamps or other marginal lands, figure the amount of land in acres that is to be

Photo by Judith Hoffman

actually farmed and divide by 20. That should give you the number of horses you need. If you have only 20 to 30 acres, you should consider two horses as this will give you greater flexibility. If you have 40 or 50 acres, you might consider three horses as you may occasionally have a need for an odd horse plowing or as a replacement for a temporary illness. The difference of a few days during bad weather can often be the deciding factor between a good crop and a poor one. A third horse and a bigger implement can make that difference.

The economical use of horses for power is a question that must be considered in connection with the operation and management of the entire farm. The very choice of the use of horses must be made carefully. Some people are not suited for day-in day-out responsibility to livestock, and work horses must be cared for before and after the long day's hard work. Some people are intimidated by the size and strength of the horse. Many can get over this, but for those who are forever "afraid" of horses, they have no place working them. The person who would depend upon work horses as a source of power must be of strong character, patient, persistent and sensitive.

Whether or not work horses are economically efficient for the small farm depends on the farm and the farmer, or in other words, the system and its engineer.

North Dakota. Photo by Fuller Sheldon

APPROXIMATE GRAIN AND HAY REQUIREMENTS FOR A HORSE WHEN NOT ON PASTURE

(1300 to 1400 pound horse)

Period	Daily ration of —	
	Grain lbs.	Hay lbs.
Maintenance (during winter)	6 to 7	15 to 17*
Light work (light hauling and miscellaneous farm work)	13 to 14	13 to 14
Medium work (cultivating corn, etc.).	14¼ to 15½	16¼ to 17½
Heavy work (plowing, discing, etc.).	16¼ to 17½	16¼ to 17½

*Partly unsalable roughages

ACTUAL CASH OUTLAY WITH BREEDING UNIT

(6 mares, 1 stallion)

Item	7 head amount	Per horse	Approx. %	Notes
Feed and bedding	0 (2042.74 value)	0 (291.82 value)	0	Assuming liberal pasture & homegrown feeds
Chores	0	0	0	Assuming family-performed labor
Depreciation	0	0	0	Assuming raising replacements & sale of horses
Interest	0	0	0	Assuming no amortization
Stabling	105.00	15.00	14%	Assuming existing structure maintenance costs
Harness	70.00	10.00	10%	Maintenance & replacement
Hoof care	315.00	45.00	40%	Assuming only corrective shoeing and trimming
Misc. vet	280.00	40.00	36%	Increased to allow for % of foaling difficulty
Gross total	770.00	110.00	100%	
Offset —Manure credit @ 160 lbs. nitrogen	308.00	44.00		

Feeding stuff	Digestible protein (%)	Digestible nutrients (%)
Dry Roughages		
Alfalfa hay, all analyses	10.5	50.3
Alfalfa leaf meal, good	16.1	56.7
Alfalfa leaves	17.4	57.9
Alfalfa meal, good	11.8	53.6
Alfalfa & bromegrass hay	7.2	46.8
Alfalfa & timothy hay	6.6	49.1
Barley hay	4.0	51.9
Barley straw	0.7	42.2
Bean pods, field, dry	3.5	50.3
Beggarweed hay	10.6	47.7
Bermuda grass hay	3.7	44.3
Birdsfoot trefoil hay	9.5	51.1
Bluegrass hay, Canada	2.8	53.3
Bluegrass hay, Kentucky	4.8	54.8
Bluestem hay	2.5	48.2
Bromegrass hay, all analyses	5.0	48.9
Broom corn stover	0.7	45.5
Buckwheat hulls	0.2	13.9
Buckwheat straw	1.2	37.5
Buffalo grass hay	3.7	47.7
Bunchgrass hay	2.7	48.7
Clover hay, alsike	8.1	53.2
Clover hay, crimson	9.8	48.9
Clover, ladino, & grass hay	11.1	53.3
Clover hay, mammoth red	6.8	52.0
Clover hay, red	7.1	52.2
Clover hay, white	10.5	55.6
Clover straw, crimson	3.8	40.0
Clover & timothy hay, 30-50% clover	4.8	51.2
Corn cobs, ground	0	45.7
Corn fodder, well eared, very dry	3.8	58.8
Corn husks, dried	0.4	38.8
Corn leaves, dried	3.5	49.8
Corn stalks, dried	0.8	40.7
Corn stover, very dry	2.1	51.9
Cottonseed hulls	0	43.7
Cowpea hay	12.3	51.4
Crabgrass hay	3.5	47.2
Fescue hay, meadow	3.7	52.7
Flat pea hay	18.4	59.5
Flax straw	5.8	38.1
Grass hay, mixed, eastern states	3.5	51.7
Lespedeza hay, annual	6.4	47.5
Millet hay, hog millet, or prose	5.6	50.7
Milo fodder	3.0	51.1

Feeding stuff	Digestible protein (%)	Digestible nutrients (%)
(continued)		
Native hay, western mt. states	4.9	52.0
Oat hay	4.9	47.3
Oat grass hay, tall	3.4	47.4
Orchard grass hay, early cut	3.9	47.8
Pasture grasses & clovers, mixed from closely grazed, fertile pasture, dried	15.0	66.7
Pasture grass, western plains growing, dried	8.6	66.5
Pasture grass, western plains mature, dried	1.3	47.1
Pasture grass & other forage on western mt. ranges, spring, dried	12.6	67.4
Pea hay, field	10.6	55.1
Peas & oat hay	8.6	52.9
Peanut hay, without nuts, good	6.6	51.9
Peanut hay, with nuts	10.2	71.6
Prairie hay, western, good	2.1	49.6
Rye grass hay, perennial	4.7	52.5
Rye grass hay, native western	3.3	52.2
Soybean hay, good, all analyses	9.6	49.0
Soybean & sudan grass hay, chiefly sudan	3.6	50.8
Sudan grass hay, all analyses	4.3	48.5
Timothy hay, all analyses	2.9	48.9
Vetch hay, common	10.1	55.3
Vetch hay, hairy	15.2	57.1
Vetch & oat hay, over ½ vetch	8.4	52.7
Vetch & wheat hay, cut early	11.4	58.0
Wheat hay	3.3	46.7
Wheat straw	0.3	40.6
Wheat grass hay, slender	4.6	51.2
Concentrates		
Barley, common, not incl. Pac. coast states	10.0	77.7
Barley, Pacific coast states	6.9	78.7
Buckwheat feed, good grade	11.7	52.5
Corn, dent, grade no. 1	6.8	82.0
Corn, dent, grade no. 5	6.1	74.0
Corn, flint	7.5	83.4
Corn ears, incl. kernels & cob	5.3	73.2
Corn & oat feed, good grade	9.2	78.5
Cottonseed, whole	17.1	90.8
Flaxseed	21.8	108.3
Molasses, cane or blackstrap	0	54.0
Oat kernels, without hulls (oat groats)	14.7	92.0

(continued)

Feeding stuff	Digestible protein (%)	Digestible nutrients (%)
Oat meal, feeding, or rolled oats, without hulls	14.4	91.4
Oat middlings	12.7	86.6
Oat mill feed	3.7	37.6
Oat mill feed, poor grade	1.4	32.3
Oat mill feed, with molasses	3.6	37.2
Oats, not incl. Pac. coast states	9.4	70.2
Oats, Pacific coast states	7.0	72.2
Oats, hull-less	13.9	89.4
Oats, lightweight	8.5	60.1
Oats, wild	9.1	53.9
Soybean seed	33.7	87.6

Feeding stuff	Digestible protein (%)	Digestible nutrients (%)
Soybean mill feed, chiefly hulls	7.8	40.1
Sunflower seed	13.9	76.3
Sunflower seed, hulled	25.2	116.1
Wheat, average of all types	11.1	80.0
Wheat, hard spring, chiefly northern plains states	13.3	80.7
Wheat, soft winter, Mississippi valley & eastward	8.6	80.1
Wheat, soft, Pac. coast states	8.3	79.9
Wheat bran, all analyses	13.7	67.2
Wheat, mixed feed, all analyses	14.3	70.6
Wheat screenings, good grade	10.0	68.7

NUMBER OF DAYS WORKED
COST PER DAY

	Cost per horse	50 days	65 days	70 days	90 days	100 days	125 days
A	746.52	14.93	11.48	10.66	8.29	7.46	5.97
B	401.82	8.03	6.18	5.74	4.46	4.01	3.21
C	110.00	1.69	1.69	1.57	1.22	1.10	.88

HORSE COST OF VARIOUS JOBS PERFORMED PER ACRE AND PER 10 HOUR DAY
(using 723 hours per horse per year of work performed)
The cost variables are shown in A ($746.52 per horse year),
B ($401.82 per horse year), and C ($110.00 per horse year)

	A per acre	A per day	B per acre	B per day	C per acre	C per day
2 horses—walking plow (2 acres/day)	10.30	20.60	5.50	11.00	1.50	3.00
3 horses—riding plow (3 acres/day)	10.30	30.90	5.50	16.50	1.50	4.50
4 horses—gang plow (4 acres/day)	10.30	41.20	5.50	22.00	1.50	6.00
5 horses—gang plow (5 acres/day)	10.30	51.50	5.50	27.50	1.50	7.50
6 horses—gang plow (6 acres/day)	10.30	61.80	5.50	33.00	1.50	9.00
2 horses—disc/harrow (10 acres/day)	2.06	20.60	1.10	11.00	.30	3.00
3 horses—disc/harrow (15 acres/day)	2.06	30.90	1.10	16.50	.30	4.50
4 horses—disc/harrow (20 acres/day)	2.06	41.20	1.10	22.00	.30	6.00
5 horses—disc/harrow (25 acres/day)	2.06	51.50	1.10	27.50	.30	7.50
6 horses—disc/harrow (30 acres/day)	2.06	61.80	1.10	33.00	.30	9.00
2 horses drilling grain (10 acres/day)	2.06	20.60	1.10	11.00	.30	3.00
2 horses planting corn (10 acres/day)	2.06	20.60	1.10	11.00	.30	3.00
2 horses cultivating (5 acres/day)	4.12	20.60	2.20	11.00	.60	3.00
2 horses mowing hay (10 acres/day)	2.06	20.60	1.10	11.00	.30	3.00
2 horses raking hay (20 acres/day)	1.03	20.60	.55	11.00	.15	3.00

723 hours — A = $1.03/hour
               B = .55/hour
               C = .15/hour

723 hours — A = $10.30/day
               B = 5.50/day
               C = 1.50/day

COMPARISONS IN COST			
The cost of two horses mowing hay (10 hour day) on different farms where the total hours that horses are used varies			
A, B, C represent management cost variables used in Tables I, II and III			
Work performed per horse per year	Cost per day		
	A	B	C
Farm no. 1 500 hours	29.80	16.00	4.40
Farm no. 2 723 hours	20.60	11.00	3.00
Farm no. 3 1000 hours	15.80	8.00	2.20

COST PER HOUR PER HORSE ON FARMS WITH VARIOUS HOURS OF USE				
	Cost per horse	500 hours	723 hours	1000 hours
A	746.52	1.49	1.03	.74
B	401.82	.80	.55	.40
C	110.00	.22	.15	.11

Above: The late Alan Conder with his six abreast of ponies on a roller.

Below: Ray Drongesen with King and Ruby.
Photo by Christene George

Item	Amount (cash value for six head horses)	Cost per horse	% of total cost	Notes
Feed and bedding	1750.92	291.82	39	Assuming no pasture, all feed purchased
Chores	1049.40	174.90	23½	Assuming all chores hired out
Depreciation	480.00	80.00	11	Assuming all geldings with 12 yr. working life (value $1,000 each)
Interest	550.80	91.80	12	Assuming amortization
Stabling	120.00	20.00	2½	Assuming maintenance of existing structure
Harness	93.00	15.50	2	Assuming maintenance & replacement of existing harness
Hoof care	270.00	45.00	6	Assuming corrective shoeing & trimming
Misc. vet	165.00	27.50	4	Worming & misc. vet supplies
Gross total	4479.12	746.52	100	
Offset — Manure credit @ 160 lbs. nitrogen per year per horse	246.00	44.00		

Notes: Feed computed at 2624 lbs. oats per year, 1.3 tons good quality hay per year and 1.8 tons of straw per year. Chores at 66 hours per head per year @ $2.65 per hour. Interest at 9%.

Dale Esgate yards out a log in the western Oregon forest. Photo by Nancy Roberts

A scatter rake in operation at the Ruby Ranch in the Big Hole of Montana. Photograph by Kristi Gilman-Miller.

A fine Belgian mule team on a walking plow at HP Days 2003. Photo by Lynn Miller.

CHAPTER FOURTEEN
VALUE OF HORSES
AS POWER

Photo by Kristi Gilman-Miller.

In March, 2000, I was invited to share some thoughts at the New York Draft Horse Short Course at Cornell University. My assigned topic was "New People Getting Started & The Future of Draft Horses". What follows are remarks prepared for that presentation, remarks which I do feel go to the heart of the subject of this chapter.

Whether or not to use horses in harness as a motive power source?

Even more than the question of whether or not to show and/or breed, this question of work choice is central to any discussion about the future of draft horses. The breeding of purebred draft horse stock, pleasure use of draft horses, and the draft horse show world can survive somewhat insulated from the work horse and mule circles. But not without serious consequences to numbers. The work horse and mule world can, in turn, survive alone but there would be negative consequences to basic questions of genetics and diversity of opportunity.

The two opposing halves, and they do oppose often, of this small draft horse industry need each other. Together they are greater than the sum of two. Together, even if begrudged and awkward, they create new opportunity for each other.

Within our measured influence, and tied to new people getting started, the future of Draft Horses is very bright. But we need to bring the work world into sync with the show/breed world.

The distillation of thought on this subject might

Master teamster and wonderful teacher, John Male of Ontario Canada. Photo by Kristi Gilman-Miller.

result in these two obvious points;

1.) The future of draft horses will depend on the aggregate total of any and all demand for them.

(In other words, if an ever growing number of various people want draft horses for a variety of purposes the future of the industry is guaranteed or at least assured.)

2.) The demand for draft horses is also assured if the industry does a good job of assisting newcomers.

(Phrased another way, the growth and stability of demand is directly proportionate to the number of new people who are successful in their beginnings with draft horses.)

Inside these points are questions which may benefit from evaluation.

1. Do we hurt our industry when we allow discussions of "why" use horses in harness to center primarily on *practicality* issues?

2. Is there a value to the draft horse industry from each newcomer?

3. Does the present day interest in draft horses and draft mules constitute a fad?

4. Does a swelling of the numbers of newcomers to the ranks of draft horse teamsters in any way constitute a threat?

5. What are the primary aspects of a predictably steady demand for draft animals?

6. Are there ways to positively affect market demand?

7. Does the industry benefit from the education of teamsters about different approaches to handling animals?

On the first question: Practicality is, especially

today, more subjective than objective.

why use horses or mules?

I have long argued the practical merits of the work horse or mule in harness. I am now beginning to wonder if that has been a mistake. Not because I have any change in my absolute belief in the pre-eminent practicality of the system but rather because it may **not** be the best way to sell the notion, to advocate the idea. This relates specifically to my discussion of Draft Horse Futures and beginners.

There are people who will quickly dismiss the notion of "working horses" as ludicrous. They're in our midst. They number among our friends and our families and our teachers and our bankers. They are offended by the idea of work horse practicality and feel a missionary zeal as they lash out against the notion.

We must address this even if understanding their motive escapes us.

We must not just dismiss or ignore this sentiment. It is part of an atmosphere in which we work and it can be most unpleasant and limiting. My suggestion is that we change the rules of the game.

If we succeed in changing the game, bringing it home, coming up with new rules that truly suit us, perhaps we may see some helpful change, if by default. Perhaps our detractors will become forgiving, understanding, or even admiring? But more important, by changing the argument we do take damaging pressure off our (the) inevitable choice.

Today in North America the choice to employ horses or mules in harness as motive power is just that - a choice. We have the choice. Thank God for the sake of the horses that we have the choice.

Making that choice means accepting the care of dependent creatures requiring feed, grooming, attention, occasional medical treatment etc.

Making that choice means accepting the care and maintenance of harness and equipment the condition of which may jeopardize the safety of you and your animals.

Making the choice means you will come to embrace a pace and rhythm to your working life that mirrors or parallels that of your working partners.

It is either that or quit.

This is the way the choice measures up:

• If you don't like draft horses and mules you have no place having them.

• If you are stuck on being thoroughly hip, cool, fashionable, admirable or envied, you have no place

Photo by Nancy Roberts.

having them.

• If you are attracted to speed and the biggest numbers, horses will never work for you.

• If you are sensitive to how your neighbors see you, horses might not be your cup of tea.

• If you believe in the absolute supremacy of specialization, horses aren't for you.

• If you are crippled by logical arguments, horses may not be for you.

I know from experience that some of you who read this hold some anxiety about this issue. You're hooked on the draft animals but troubled by discussions which suggest, or outright claim, that working these animals in harness can be a viable practical option for farms and woodlots. You are certain that such claims are indefensible and set the industry in a silly light. These are the right places and this is the right time to look at this difficulty and this sentiment. So before some of you decide you don't want to hear any more, I need to make some bold and defensible statements.

You may be surprised to learn that this entity we call a draft horse industry accounts for hundreds of millions of dollars in gross sales throughout North America each year. This industry supports many peripheral small businesses such as our own publishing company.

It is my business to constantly measure this industry and its potential market universe. I have done this for 25 years. It is my contention that this industry could double in size within a two year time frame if certain social impediments were addressed and/or erased.

A doubling of the size and breadth of this industry.

But it won't happen without change. It can't.

These are the obstacles as I see them.

Many, not all, old hands (anyone with 6 or more years experience?) with draft horses treat the starry-eyed newcomers with mockery, disdain, ridicule, distance and even sometimes disgust. It troubles me to have to point out that the truly helpful old hands, though they are responsible for most successful beginnings with horses, aren't enough in number.

Many, not all, horsefarmers and horseloggers look down their noses at people who only breed and show their draft animals. I know if I look inside that I am guilty of this myself and it is destructive.

Many, not all, show and breed people look down their noses at people who choose to work their animals. What a shame. A powerful market alliance would be possible, as it was in the not so distant past, if such animosity didn't exist.

Many established draft horse folks feel they have no obligation to help newcomers. They hurt themselves with this attitude.

Many are the newcomers who by virtue of enthusiasm alone were ready to spend the money to buy animals and equipment, but came up against a wall. And it never happened or sputtered as a misguided first effort. With each of those failed dreams and beginnings each of us lost out.

This is not a call for establishing any centralized organization. We have seen those efforts and they do not work.

This is a call for reflection and preparation.

Reflection on how we see ourselves. And *preparation* for the massive changes which are on the way as we speak.

Through the offices of *Small Farmer's Journal*,

we hear from and about an increasing number of successful business people who've made the decision to get involved with draft animals. Some, of course, are drawn to the show and breeding world. But what interests us for this discussion is that growing number of professionals (lawyers, pilots, doctors, etc. etc.) who have decided they want to add the working of horses and mules to their world. These are people to whom the expense is easily affordable. These are people who, through successful independent lives, are accustomed to getting things done. They are accustomed to getting answers, acquiring what they feel they need, and making a mark with whatever they do.

I hear so often from this group that the prices of our animals and equipment seems to them so low. They find easy justification for double to triple the amounts most of us pay.

I hear so often from these folks that they feel many in our industry look down on them and they don't appreciate it. They feel many in our industry are obviously poised to take advantage of them if possible. And this observation, so easily made, humors them.

For many of these people their interest in the draft animals is genuine and long term. They will succeed even if it must be done without us. These are people who plan on having an impact on the draft horse industry.

I'm afraid if we don't pay attention, we'll be left behind. We need to use our smarts.

****

Picture a view of four or five adjoining small fields. None of which are larger than ten acres. All of the fields are neatly fenced. The fencerows include trees and bushes and hedge plants. In one field livestock are grazing. In another, plowed ground mellows. One field holds ripening grains. Yet another contains a maturing hay crop. There is perhaps a fifth field of vegetable rows needing cultivation.

The sun is shining but the horizon shows an approaching rain storm.

The day's field work plans are determined by the weather and the readiness of soil and crops. Unless a third element should have an influence.

Power Source.

What are the choices, what are the available options, what adds to the picture, what contributes to the balance, what is right for the individual farmer? What power source will sensitize the farm's landscape

and the farmer's purpose? What's the best way to get the work done? Can the work get done and the poetry and life affirming beauty of the scene be protected, enhanced, showcased?

....

Stop! Where's the return on investment? Where's the profitability?

But the picture is either incomplete or sanitized by a dreamer's reality vacuum. And we allow that incompletion because no one stops to question the question. If a farmer wants (perhaps needs) to paint work animals into his dream farmscape REGARD-LESS and IN SPITE OF practicality issues - I want to strongly suggest that we're all more comfortable with it.

When we insist on practicality and reasonable motives we crowd each other in curious and dangerous ways.

Our post-industrial twenty-first century world rushes to deny and squash that dreamer's vacuum, that dreamer's frailty, for fear it might defrost the temporal sanctity of progress.

Paint a simple picture for yourself of how you would have your farm be. Allow that unfolding dreamscape to suggest options, choices. Hold at bay that portion of the 'outside' world which screams "you are no better than the rest of us and we're heading this way. Come on." Only the lonely, the intrepid, the brave, the innocent, the blind, the passionate, the sad, the courageous, the simple-minded, the dreamer, the dedicated, the unsophisticated, the invalid, the unrea-sonable, the introspective, will go this route. It is my contention that this route leads more often than not to personal fulfillment.

I don't work horses because I want to. I do it because I need to. This is an important distinction and one which might lead us to that about-face I spoke of earlier. I've been working horses for thirty years. I have the skills, the animals, the farm and the paraphenalia. And I can work them whenever I wish. I don't have to work them because some outside force or urgency dictates. That was true in the beginning. I had to harness up and get to the field or the work didn't get done. I, personally, never felt the "had to" in any negative way because from the beginning I was raptur-ously in love with all of it. There is the clue. That love part.

When you are in love with someone all practical concerns fall away. You have to be with that person because you must, you need it.

Same thing with objects and endeavors - if you

love them you must be with them. It is a passion and you will discover or create a justification if you must. But I say passion is good enough.

We all understand it. Wouldn't it be easier to sell the idea of work horses and mules as passion? Even though deep inside we know they are practical in amazing ways.

That second question: Is each newcomer significant? Most definitely. When someone comes to us for help, no matter how elementary - how silly, we must pull out the stops and help. WE MUST. Each successful beginning has a quantum effect of our industry. Animals harness, instruction, club membership, equipment, publications, all get bought. You'd think everyone would agree. Not so.

There are draft horse organizations that require that new members be sponsored and voted on secretly. New people are regularly denied membership. There are draft horse shows which are by invitation only, no uninvited newcomers are allowed to compete.

The net effect of this attitude is in my opinion putting a lid on draft horse prices and overall growth.

In 1978 I spent $2500 each for two good registered Belgian Mares. Today I can repeat that purchase. No appreciation in value. That won't last, values will skyrocket with or without us. And the key is newcomers.

Exclusivity in our draft horse community will bottle up potential growth, and when it explodes we may be left out.

Third question: Is it a fad? No, that quality in us which attracts us to these animals and to the work will never fade.

Fourth question: Are new people a threat? Absolutely not. And probably yes.

Those new professional people I spoke of earlier; they are the ones who will some times drive the old hands nuts by becoming instant experts. After six months of successfully driving a team they sometimes see themselves as the new torch bearers. And they are a genuine threat because they have the past life experience and the wealth to do just about whatever they want. They can buy up an entire gene pool of draft horses, they can start up a new draft horse show, they can begin a brand new draft horse magazine, they can move to corner whatever piece of this industry they might want. And they will. Unless we wake up and accept the challenge they represent.

Good competition should wake us up. Wake us up to possibilities. These people will be less of a threat

and more of an asset to us if we are there first, by their side, with help and real friendship. We create the vacuum they would rush to fill by ignoring them and denying them access to this established little industry.

Fifth and Sixth questions. Steady demand will be the result of improved animal quality along with those previous points about helping new people. We need to be thinking about value-added product. Trained animals, animals sold with harness, bred animals, 'reputation' animals. We need to do follow-ups with people who've bought our animals. We need to work on connectivity, repeat business. Perhaps we should be selling horses with operating manuals? Perhaps we should be selling bred mares with a guaranteed buy-back of the foal?

I have been as guilty as many others of refusing to sell my horses to novices when what I should have been doing was giving an extra measure of access to these people. What better candidate for one of your horses than someone you've helped to train, than someone who will call you when there is some new challenge and before there is a problem?

Draft Horse organizations might be well served to consider expanding events to be more inclusive, combining plowing matches with swap meets and clinics just as one example.

Seventh question: New ideas about handling animals. The more trustworthy animals our industry puts out the better the beginner experiences and the faster the growth. New ideas and new approaches to training are very important.

Our modern times of synthetics and artificiality cry out for moments of comfort which cannot be purchased but must be earned through the experience of them. When a string of moments of comfort can be combined with actual working accomplishments (i.e. the hay is mowed and raked or the field is plowed or the logs are skidded to the landing) the people involved are invariably improved. Love what you do. Feel good about what you've done. Simple solutions for a dark time.

the $ value of the work horse

For the sake of clarity I need to point out that my measurement of the cost-value of working horses is based on the model of a mixed crop and livestock farm where horses are used as the PRIMARY power source. And that fertile mares are used, at least in part. And that foals are raised for replacement and sale. Let's use the unit model of two 3-year-old draft mares ready to go to work. For the sake of discussion, assume with me that they work for you at least 12 years to the age of 15 (knowing they may do far more) and each foals six times with five foals each surviving (this is not to suggest any average but rather a conservative assumption). These mares will produce seven tons each per year of manure for fertilizing use. And may easily provide 800+ hours (five full time months per year) of labor in harness as a power source. Plus, using them instead of a tractor system will save money.

I have computed that in order to run our ranch with tractors rather than horses it would cost me a minimum of $25,000 per year (and possibly as high as 40,000) to cover mortgage payments on the tractors and implements, fuel/oil, parts/repairs and depreciation (or replacement cost). This is with used equipment and doing most of our own repairs. We use more horses than we need for a variety of reasons, but long experience has told us we can do our work handsomely with three teams. (It may take you less horses or more.) If you factor seven horses so that you have a replacement in case of injury and divide that $25,000 by 7 you come up with a gross cost savings per horse of $3,570 per year (at $40,000 cost, it would be $5,700). We will round out the estimated savings over tractors to $4,600 per horse per year. For a team, this comes to $9,200 per year or $110,400 saved over 12 years!

Since these mares will be producing their own replacements, this offsets the initial cost of the team. The remaining four foals might be sold for anything from $600 to $2,000 each (average $1,300 x 4 equals $5,200).

The labor of one horse in the farm field is worth at least $2.00 per hour if you look at the hard cost of replacing same with human labor or internal combustion bought/leased or rented. 9,600 hours (12 x 800) of labor from one mare would then be worth $19,200 or $38,400 for a team. (This is a separate line item from the savings over tractor systems as it does not take into effect related required implements, tools and expertise

tied to tractor systems as carrying their own cost.)

The manure produced comes to 84 tons in 12 years at $20 per ton for the nitrogen value or a total of $1,680 per horse with $3,360 per team.

There are intangible and less tangible benefits from the working mares which are difficult or impossible to quantify (i.e. advertising value, attraction, satisfaction etc.). Using the tangible figures, we come up with a total benefit over 12 years of $26,080 per mare or $52,160 for the team.

Here's how it might stack up:
twelve year value of a pair of working draft mares

foals crop (less replacement pair)	*$5,200*
value of labor	*$38,400*
manure crop	*$3,360*
savings realized over tractor system	*$110,400*
total value	*$157,360*

Outside of the foal crop, none of this represents actual dollars received from sales. But to my mind, savings of over $150,000 over 12 years translates to more money staying home. Simply put, that working mare is worth $5,000+ per year to you and me. Farming with horses does allow you to keep more of what you make.

You interrupt me at this point and ask, "Hey, what about the costs of keeping those mares. They ain't free, you know."

Pretty close to free in my estimation. If you went out and bought a tractor for $3,000 and put a couple hundred a year into its upkeep and got back $5,000+ per year for a dozen years without letup you'd be jumping for joy and hoping the IRS never figured it out.

Okay, let's look hard at the inputs:

costs

The single most important input with cost is working experience. In other words, BEFORE you can make working horses a comfortable reliable operating scheme for your farm, you must acquire the necessary skill level. It comes from watching, learning, experience and curiosity. And acquiring the skill of a full-fledged farm teamster will cost – time, sweat, adversity, anxiety, and perhaps money. Remember, if you owned a team of young draft mares magically – tomorrow – you would not be able to realize their full value to you without your first having the skill to use them properly. This is the single biggest reason we don't have a flood of new horse farmers all across the countryside. It's not because it's impractical. It's because it takes commitment and work to develop skill and craftsmanship. But once you have, it is yours for life.

Here's a secondary list of somewhat regular expenses to working horses which includes:

worming
hoof care
veterinarian attention
breeding fees
and miscellaneous supplies

The worming costs us $50 per horse per year. Hoof care we do for ourselves (they go barefoot). If professionally done, it could cost $100 per horse per year for trimming (considerably more for shoeing). Vet care can vary wildly, so for this discussion we averaged our expense over ten years and it came to $150 per year per horse. Breeding fees can cost $200 or much more per mare per year unless you keep a stallion as we do. Miscellaneous supplies will eat up $50 per year per horse. With each of these line items, it would be easy to spend less or more.

Stacking up these secondary expenses to match how we ran the values for two mares for 12 years;

worming ($50 x 2)	$100
hoof care ($100 x 2)	$200
vet care ($150 x 2)	$300
breeding fees ($200 x 2)	$400
misc. supplies ($50 x 2)	$100
year's total	$1,100
twelve year total	$13,200

(Note: by doing your own hoof care and keeping a stallion you can save $600 a year)

If you must buy your own feed, add in your own costs.

All of our other expenses are minimal. We raise our own hay. You might quickly say "but you could sell it so you need to record that lost income as expense." We do not sell hay and we will not. It is intrinsic to our farming philosophy and methods to feed that hay to our own livestock and have the manure it has produced returned to the soil. In this way we are not selling our fertility. When our workhorses consume our hay and return us manure, they are in effect working as a component of our farming system. We choose to see it this way; we grow the forage to feed the livestock so that they may produce manure to feed our soil and produce a better crop. The degree to which we have

surplus calories to market is in direct proportion to how successful this cycle is. On our balance sheet it costs us nothing to feed a horse. Rather it costs us to produce forage which is converted to a higher value product, manure which is in turn converted to an even higher value result, fertile soil. Whereas such measurement may not satisfy an accountant or a banker it is critical to our decision-making process; it affects how we value our animals and why we choose to retain fertility at all costs.

You on the other hand might reasonably say, "We have to buy hay and grain for our horses, so it is a cost to us." That may well be your circumstance but I believe that in such a scenario you may NEVER realize the full value of the work horse. It is, in my estimation, important for you to "allow" the practical realities of the working team to have a shaping effect on your operations. For example, rather than seeing the speed of horses as a limiting factor, I suggest that you allow for it and set up operations to spread the work more evenly over the whole year. In this way you will have less or no problem getting it all done and the resultant "shape" of your operation will, doubtless, benefit not only the horses and you but also the whole of your farming. I'm sure there must be exceptions but I still do believe that any mixed farming or ranching operation which employs work horses must produce forage in some form (even waste) for those animals. Not to do so is to lose an important economic component. I know it is still fashionable to "specialize" and focus on one thing and do it well. But it's a mistake, certainly with farming. Farming screams for diversity and most especially if work horses or mules or oxen are to be part of the picture.

I knew a successful rancher who read that he needed to look hard at the costs of his operation and be prepared to make hard choices if he was to "get modern and be legitimate." He did this. And his first decision was a big one. He decided to specialize and just do cattle. He figured that he could not afford to raise his own hay any longer. He decided to buy more cattle and pasture his hay ground. He sold all his loose hay making equipment. And he sold his feed teams and sled and hydra fork. He contracted with a local rancher to buy big round bales which he counted on feeding out with a new attachment for his tractor. Things didn't turn out as they were supposed to. After two years, the price of hay skyrocketed and the price of cattle dropped through the floor. He did another hard fiscal examination and decided to sell all of his cattle thereby cutting his future losses. This decision was helped along by an offer from a neighbor to lease all the ranch ground for grazing. So he signed a three-year lease and took a job for day wages riding for a big ranch down the road. And while sitting in a drafty line camp cabin that winter, a long ways from his wife and kids and ranch, he had plenty of time to think about the fact that he had worked awfully hard those first years to get and build up his little ranch. He had once been mighty proud of the skills he practiced and practiced well. And it just didn't seem fair that by being "legitimate" and "modern" he had given it all up. He'd penciled it all out until his efficiency was 100% and he had "down-sized" himself into a backroom.

Back once again to the "value of work horses" discussion: In order for any teamster to realize the full value we suggest from his or her horses it is paramount that all these concerns be met:

· YOU must know how to do the work required.

· You must have well-trained, well-behaved, healthy fertile mares.

· You must actually need to work your horses and then do so.

· You must minimize all health concerns.

· You must raise and sell a minimum number of foals.

CONCLUSION

When I did these calculations I was surprised to see the strong numbers. I knew the horses were important to me but I never wanted or needed to quantify that. You see, I love working my horses and that was enough for me. It is, after all, WHY I do it. I do it because I love it. Not because the numbers look good.

Side note: In North America, ours is the most advanced animal-powered technology the world has ever seen. World War Two derailed us from exporting that technology. Soon we may, if we choose, take up where we left off. Norway, China, Thailand, South Africa, Paraguay are all waiting.

The value of what we know about working horses is current worldwide. And growth potential in those other countries is staggering.

"I don't work horses because I want to. I do it because I need to."

A white mule and carriage in the French Quarter of New Orleans.
Photo by Lynn Miller.

CHAPTER FIFTEEN
WORK YOU CAN DO WITH HORSES

Walt Bernard of Dorena, Oregon asked this visiting group of school children to pull the same walking plow his two horses were walking away with.

The work which horses can do depends on their weight, their muscle tone, and their endurance. At steady and continuous work for 10 hours a day, the pull (or draft) for the animal should not be more than one-tenth of its weight. For example, a 2,000 lb. horse should not be required to exert an average constant pull of more than 200 lbs. (to 250). For a brief moment a well-trained, well-conditioned horse can pull 10 times (plus) the normal rate, exerting a pull sometimes greater than his own weight. When we speak here of the "pull" we are not speaking of the weight of the load but the actual weight of the drag. If you had a sack of feed and hooked a spring weigh scale to that sack and pulled – the pounds registered on the scale when the sack first moves will be the "pull." Obviously if the same sack is put on a wagon that rolls easily, it will require fewer pounds of pull to start the load. A team of horses can pull several ton of weight on wheels.

In measuring the rate at which horses perform work, the unit called "horse-power" is used. This is the performance of 33,000 foot pounds of work per minute. (One foot-pound is the amount of work done in lifting one pound one foot against the force of gravity.) A horse weighing 1500 lbs. is able to work steadily at the rate of approximately one horse-power. During hard pull the

Lynn Miller drills oat seed in 1979. Photo by Christene George.

same horse might be exerting a rate of 10 hp or more.

Weight is an important factor in determining how much a horse can pull. Feed, harness and the skill of the teamster all affect how much work horses can do. (See FEEDING and HARNESS chapters.)

The kind of road surface is a big factor in determining how much weight a horse can move on wheels. While only 25 to 50 lbs. of draft are required after the load is started, to haul a load of a ton (including weight of the wagon), on a level pavement made of concrete or asphalt, the draft on a dirt or gravel road is 75 to 225 lbs. or more per ton.

On a soft surface (especially mud), the height of wheel and the width of tire are important, as they affect the depth that the wheels cut into the ground. On a good hard, smooth surface there is very little difference in draft with different kinds of wheels.

Horses are most efficient in doing a full day's work when walking at a speed of 2.5 miles per hour. However, walking as fast as 3 miles per hour does not reduce efficiency if the animal's conformation allows for ease and comfort. It is up to the teamster to make distinction as to what is the most efficient walk for the animals used.

on the farms

Farms can be operated well with one good work animal for each 25 acres, excluding pasture. Mixed farming requires at least one-fourth of a farm be in pasture. This means 4, 5, 6 or 8 work animals will do all the field work on 120, 160, 200 and 240 acre farms, respectively.

The two horse team will:
  plow 1 ½ to 2 acres per day,
  disc 10 acres per day,
  harrow 10 to 12 acres per day,
  drill 6 to 10 acres per day,
  plant 8 acres per day,
  cultivate 8 acres per day,
  mow 8 acres per day,
  rake 16 acres per day,
or do wagon work 25 miles per day.

Three horses on a farm will:
  plow 3 acres per day,
  disc 15 acres per day,
  harrow 15 to 17 acres per day,
  drill 12 to 15 acres per day,
  or cultivate (with 2 on one cultivator and 1 on another) 12 acres per day.

Four horses on a farm will:
  plow 4 acres per day,
  disc 20 acres per day,
  harrow 17 to 20 acres per day,
  drill 20 acres per day,
  plant (with 2 planters, or doubled) 16 acres per day,
  cultivate 16 acres per day,
  mow (with 2 teams) 16 acres per day,
  or rake (with 2 teams) 32 acres per day.

Six horses on a farm will:
  plow 6 acres per day,
  disc 30 acres per day,
  harrow 30 acres per day,
  drill 30 acres per day,
  plant (3 teams) 24 acres per day,
  cultivate (3 teams) 24 acres per day,
  mow (3 teams) 24 acres per day,
  or rake (3 teams) 48 acres per day.

and so on.

To calculate the amount of work done, figure speed (i.e., 2.5 mph), times day's length (time actually moving). Multiply resulting distance in feet by the width of the implement (i.e., 8 feet) and the result will be square footage covered. (Divide by square footage of an acre for acreage measure.)

A list of jobs that can be performed on the farm by horses include:

Lynn Miller planting corn with Bud and Dick.

Plowing, harrowing, springtoothing, discing, rolling, leveling, drilling seed, planting seed, spreading fertilizer, spreading manure, hauling irrigation pipe, cultivating, mowing, tedding, raking, loading, baling, hauling hay, binding grain (including corn), picking corn, chopping silage, hauling silage, digging potatoes, miscellaneous harvest work, hauling winter feed, hauling firewood, collecting maple sap, skidding miscellaneous, and more.

how many logs

It is more difficult to give even rough estimates of log skidding production with horses because there are more variables.

Depending on the length of the skid road, how hard the going is, how large the logs are, the speed of the horses and the ability of the horse logger, production may vary from one thousand to eight thousand board feet per day per team. Perhaps the most important current fact about horse logging is that, given a healthy lumber demand, the conscientious, intelligent logging operation seems to function well with horse power, it is not a get-rich situation, but it is viable.

As to how big a log can be pulled, here again the variables of weight, skid-road, horses and teamster all take effect. The teamster will "learn" his team's limitations and should do so gradually, working up to the big stuff.

in the city

Drayage work and people hauling are big new frontiers for the horse power scene. Every day a new little example pops up somewhere of people using horses in unexpected but welcome ways and places. Like in California, where community parks are being maintained with horses and horse-drawn streetcars are moving people. In Oregon, where garbage, freight and people are being hauled. In Washington, where heating oil is delivered by horse-drawn wagon. In British Columbia, where Clydesdales haul a traveling theatre group. And the list goes on and on.

The potential seems wide and new ideas plus new people will keep the doors open for the horse in harness.

The exciting aspect of the work horse is this wonderful flexibility combined with an easily enjoyed overall character. What person isn't charmed by the sight of an oncoming horse-drawn buggy or a hard-working team in the field? Add to this the FACT that the use of horses and mules in harness IS PRACTICAL if approached intelligently and sensibly. And with that you have lots of promise. There will be more equipment and information available which will in turn improve the possibilities ultimately resulting in an even greater variety of jobs that can and will be done with horses.

Jim Bowers of Port Angeles, Washington.

Jess Ross with buck rake team at Montana's Ruby Ranch. Jess is a superlative horseman and a dear personal friend. Photo by Kristi Gilman-Miller.

Tom Odegaard with six Belgians on field cultivator. Photo by Fuller Sheldon.

Bob Nygren with Percherons on team grain drill. Photo by Nancy Roberts.

Gary Eagle with five abreast on field cultivator. Photo by MaryLyn Eagle.

Below: Lynn Miller with Belgians on manure spreader.

Bob Nygren with eight foot disc. Photo by Nancy Roberts.

Paul Birdsall of Maine, horsefarmer and legendary teacher, running a corn and bean planter.

Kris Woolhouse enjoys a wagon ride with her farm team in Dorena, Oregon.

Below: A mule-powered Vis-a-Vis carriage, hauling for hire in the French Quarter of New Orleans. Photo by Lynn Miller.

Amish buggy with team in Ohio.

A crowd struggles to watch as four Percherons pull a baler and bale wagon. Photo by Lynn Miller.

A Missouri sorghum press by Bill Heincker.

Charlie Yearian, and his family, have helped hundreds of folks with their draft horse beginnings. Photo by Helen Eden.

Jack Eden, a world class mule skinner, runs six on a gang plow in the Bitterroot country of Montana. Photo by Helen Eden.

Carl Russell of Vermont loads his log scoot and hauls through the snow. Photos by Kyle Jones.

Big Rowdy and Digger seeding oats at the Fall Harvest Days in the eastern Missouri Ozark hills.

Ron Arnold plants corn on his Kentucky farm.

Bulldog Fraser, of Montana, shows off special summer field work head gear as he spreads fertilizer with four abreast.

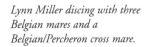

A stout Haflinger works a homemade horsepower, built out of McCormick Deering No. 7 mower gear set on its side. The pitman shaft is converted to a power shaft which, in this case, is running a large ice cream maker at the Indiana Horse Progress Days. All photos by Kristi Gilman-Miller.

Left: A Haflinger runs a modern day treadmill also powering a large ice cream maker at HP Days Indiana.

Lynn Miller discing with three Belgian mares and a Belgian/Percheron cross mare.

Above and Below: Tom Odegaard and his Belgians on a Deering binder in North Dakota. Photos by Fuller Sheldon.

A round baler and a motorized forecart in Indiana.

Logging demonstrations at HP Days, Indiana. That's Farmer Brown in the bottom pic. Photos by Kristi Gilman-Miller.

Dann Harris, of Hanover, Ontario, designed this log cart. Here Jack, Sandy and George have moved four hard maple logs.

Left: Anthony Arnold with Tom and T.T. Titan skidding logs in British Columbia. Photo by Mark Hirkila

Jarrod M. Boldt plants the same field his grandfather planted 70 years before in Shakopee, Minnesota.

A Yoder Produce built horsedrawn vegetable transplanter. Photo by Kristi Gilman-Miller.

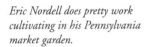

Walt Bernard adjust his straddle row cultivator near Dorena, Oregon.

Eric Nordell does pretty work cultivating in his Pennsylvania market garden.

An all woman combine hitch from circa 1920 along the Columbia River.

C. T. Fields wheat combine outfit in August of 1929.

Joseph riding Ted to and from the shocks on grandfather Joe's North Carolina farm. Photo by John Hartman.

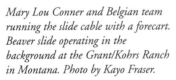

Joe Anderson and a 9' dump rake at work in North Carolina. Photo by John Hartman.

Four Spotted Drafts pull forecart and round baler in Pennsylvania. Photo by Lynn Miller.

Mary Lou Conner and Belgian team running the slide cable with a forecart. Beaver slide operating in the background at the Grant/Kohrs Ranch in Montana. Photo by Kayo Fraser.

Doug Hammill and Adie Funk with four Clydesdales on small combine harvester in Montana. Photo by Carla Hammill.

Oregon's Lise Hubbe, raking hay with her Belgian team.

Dr. Doug Hammill discing in Montana. Doug is a true friend of nearly thirty years and an associate editor of Small Farmer's Journal.

Mike McIntosh drives four Belgians on the header while father, Mac MacIntosh handles header box wagon at the Dufur Threshing Bee in the Oregon town of the same name.

Right and both below: Lise Hubbe of Scio, Oregon, one of many women proving, with a smile, that working horses is a fulfilling vocation.

Above: Today's Belgian horseman Lester Courtney of Washington, 30 years ago, was quite a hand with pony teams.

Dale Esgate at a logging show near Crow, Oregon. Photo by Nancy Roberts.

Mic Massey's Shire, Herbert, and hay wain at home farm near Norfolk, England.

Left: V plow with Fran Kueker driving Kaibab near Verndale, Minnesota.

Gary Eagle and his feed sled up in Washington's Okanogan highlands. Photo by MaryLyn Eagle.

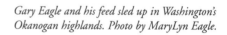

Left: Eric Nordell's toolbar cultivator on the Nordell's Pennsylvania farm.

The new Groffdale disc at work on the Nordell farm.

Rich Hotovy cultivates and fertilizes with new I & J unit at the PA Horse Progress Days. Oh, by the way, those are Fjord horses! Photo by Lynn Miller.

Four horses abreast pull a motorized forecart and haybine at PA HP Days. Photo by Lynn Miller.

A motorized baler behind a basic forecart comes ahead of the big bale wagon. Four Belgians abreast in PA.

Above: Mississippi's Kenny Russell on the mower with Lynn Miller at a Russell workshop. Kenny and Renee run one of the best workshops anywhere.

Best place to end the chapter, attentive folks at a Russell workshop. The single most important investment new people can make in this realm is not in horses nor is it in equipment. The most important investment is in the best learning.

CHAPTER SIXTEEN
EQUIPMENT

For anyone who might doubt the present and future vitality of working horses we only need point to the explosion, over these last twenty years, of new equipment design, manufacture and sale. When I wrote the original Work Horse Handbook there was little or nothing in the way of new implements. Looking through this chapter, as well as through the photos from all the rest of this book, we see a staggering assortment representing terrific ingenuity and promise. With the important help of the annual horsedrawn technology trade fair we call Horse Progress Days, there is certainty that we will see continued innovation and vitality in this arena.

forecarts

Before we had this flood of new stuff, most of us were dependent on the old original implements and the use of forecarts to pull tractor equipment.

For these last 60 years, horse farmers and loggers have come up with a vast array of different forecarts, the majority of which are based on a simple design principle: two wheels – weight balanced and slightly back – tongue coming direct from axle – and hitch point close to axle and under the seat. Carts like this have been hitched to by 2, 3, 4, or more horses, and all sorts of implements have been attached behind. The list includes harrows, discs, rollers, grain drills, plows, side-delivery rakes, wagons, manure spreaders, logs, etc. Innovative horse farmers have even hitched to balers, combines, choppers, silage wagons and more. The sky and imagination seem to be the limit.

This author, using his own experience and borrowing from others, has come up with a different principle for forecarts. As pictured on pages 318 & 319, this cart has three wheels (the front one steering), and an adjustable seat. It allows, because the seat is forward, for greater flexibility in implement hitch and there is no weight from the cart on the horse's neck. Dimensions are given on the next pages to make it possible for you to perhaps build your own or have one custom built. Some of the construction requires steel cutting and welding, which may have to be custom-done.

The cart is built on a 4" square tubing "T" frame with used Rambler wheels on the rear axle and an offset axle bracket and wheel (off the rear of a junked side-delivery rake) for the front end. The seat, including footrest, is a separate unit which slides down

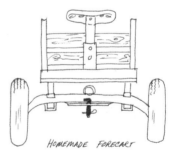

HOMEMADE FORECART

Homemade forecarts of every stripe, size and design are seen all over North America. Often farmers build them on old front truck axles which have had steering spindles welded shut, see drawing on right. Others use tube steel axles with flanges welded on to receive lug bolts.

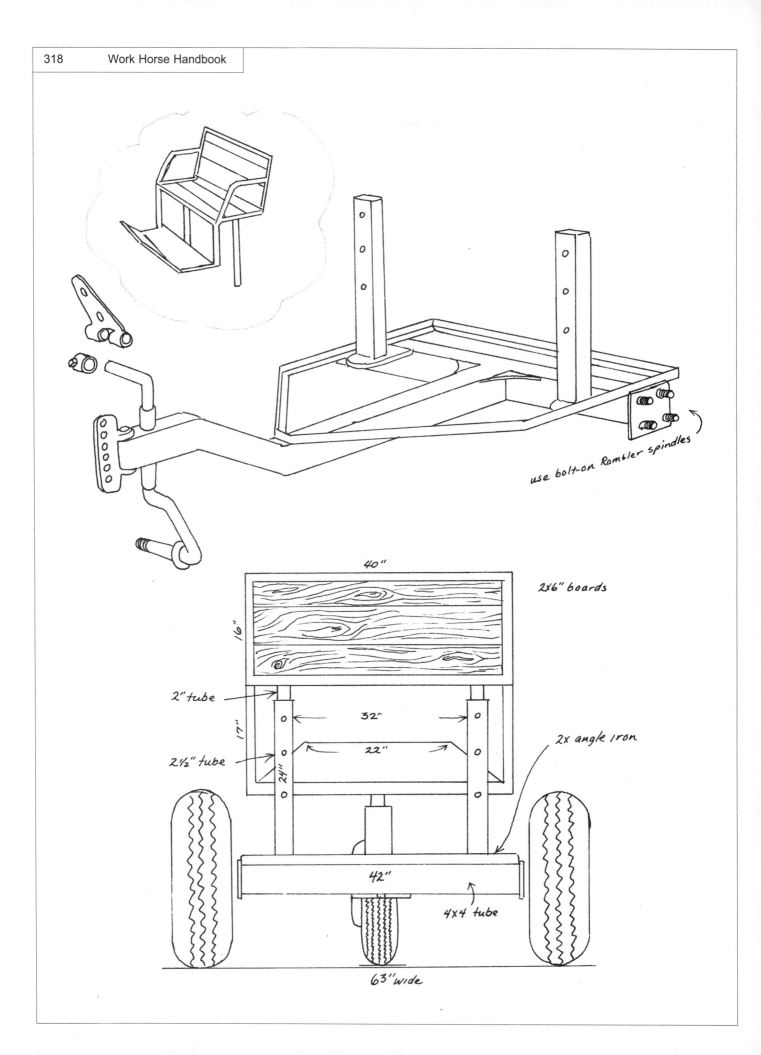

use bolt-on Rambler spindles

40"

2x6" boards

16"

2" tube

17"

2½" tube

32"

22"

24"

2x angle iron

42"

4x4 tube

63" wide

Ray Drongesen with three wheeled cart hooked to gooseneck tractor rake.

into standards, allowing for adjustable height. This works great with big hitches allowing the teamster to "see" his leaders and the rigging. The tongue slides onto the steering shaft in the same manner as a plow tongue and carries just its own weight. The hitch is adjustable up and down for different size animals and outfits. Because of the triangular shape of the body of this cart, and the steering tongue, even with four abreast hitched, this unit will turn around almost on top of itself. The frame is stout enough, and there is room enough under the seat, to handle a 5-hp motor mounted to propel a power-take-off unit and/or even a remote hydraulic system. If you add brakes this cart would be able to handle ANY tractor equipment.

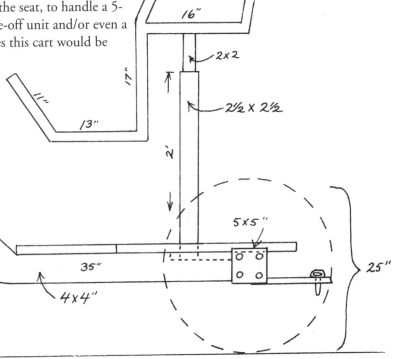

Three wheeled cart pulling spike tooth harrow.

Pioneer Equipment, of Dalton, Ohio, makes an extensive line of forecarts with a long list of options. Thousands of these carts are used on farms all over North and South America. They have proven themselves to be essential tools for the well provisioned horse farm.

Above: Scraper option for Pioneer forecart.

Another company designing and building a good line of useful forecarts is the White Horse Equipment company of Gap, PA. They build standard, hydraulic, battery powered and motorized units.

Left: A new innovation for White Horse carts features an adjustable axle allowing that the weight of the hitch may be shifted to balance the cart.

An unusual seat spring used on some White Horse carts.

sleds

Sleds, stoneboats or workboats are extremely handy units for the farm. They can be made in the farm shop, see plan next page, or purchased ready-made like this exceptional unit built by Pioneer Equipment.

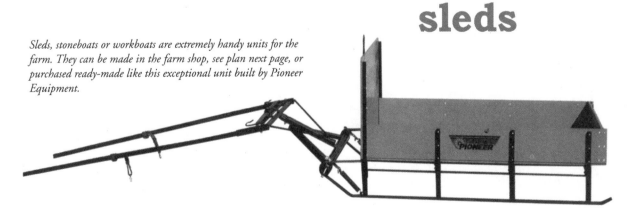

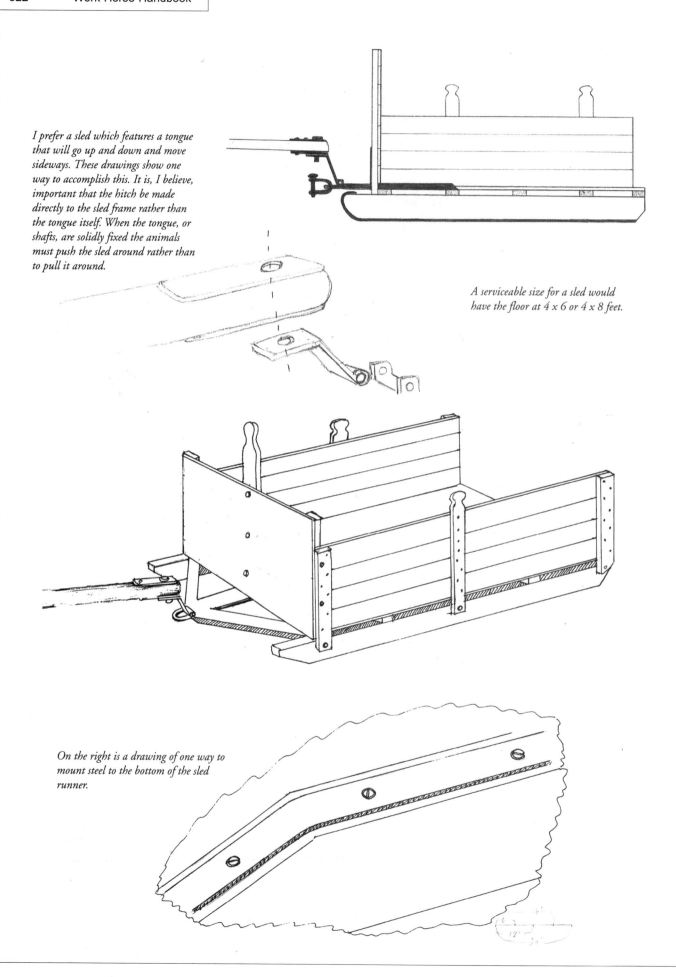

I prefer a sled which features a tongue that will go up and down and move sideways. These drawings show one way to accomplish this. It is, I believe, important that the hitch be made directly to the sled frame rather than the tongue itself. When the tongue, or shafts, are solidly fixed the animals must push the sled around rather than to pull it around.

A serviceable size for a sled would have the floor at 4 x 6 or 4 x 8 feet.

On the right is a drawing of one way to mount steel to the bottom of the sled runner.

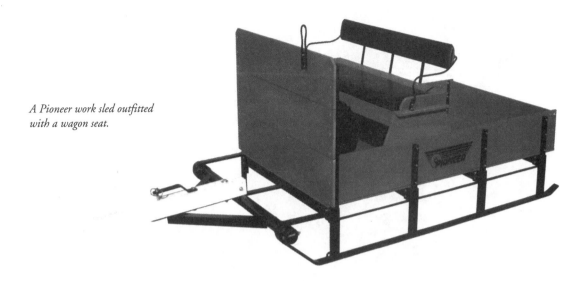

A Pioneer work sled outfitted
with a wagon seat.

Many farmers have come up with their
own designs for a ground drive PTO
operative forecart. Here are two styles
which have been shared at Horse Progress
Days.

Pioneer, Gateway, White Horse, and others all make motorized forecarts with many configurations and options. They can be expensive but they do allow that virtually any tractor tool be used with horses.

These carts can come with diesel or gas engines, battery power, or combinations. They are available all the way up to 75 HP. (Not the true HP, that other stuff.)

Don't forget, you can get the implement with its own motor and use the standard cart to good advantage.

plows

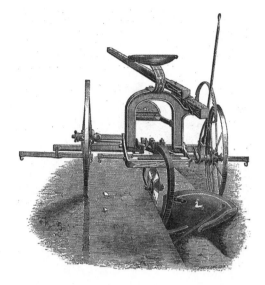

Old walking and riding plows will be with us for a long time. They are marvels of design and structure. But they can be a challenge to repair and maintain. Thank heavens we have an array of new plows which answer that concern.

Pioneer Equipment makes the stalwart walking plow in both directions and most sizes. I use one and love it.

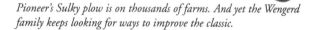

Pioneer's Sulky plow is on thousands of farms. And yet the Wengerd family keeps looking for ways to improve the classic.

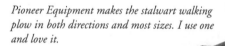

Photo by Chris Feller.

On the right: Pioneer makes a four abreast evener which puts three horses on the land when plowing with their sulky.

White Horse sulky plow and a closeup showing hitch and spring.

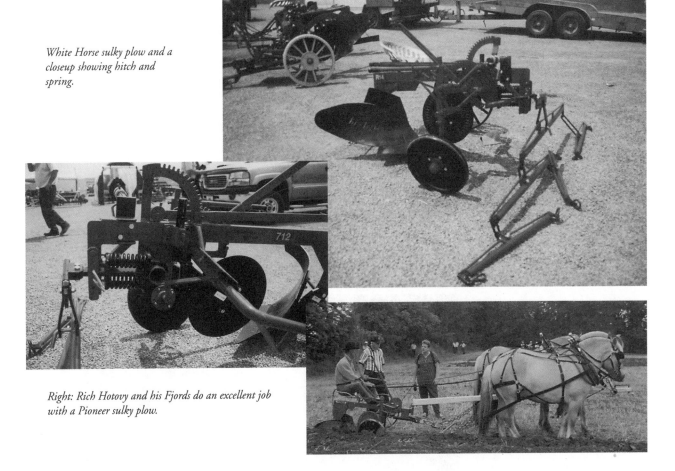

Right: Rich Hotovy and his Fjords do an excellent job with a Pioneer sulky plow.

Right: HP Days is a place to see many new implement prototypes looking for manufacturers.

The Pioneer gang plow is a marvel of engineering employing new variations on the principle of the old foot-lift plow.

White Horse Machine makes this hydraulically operated two way plow which allows that you work without dead furrows. The farmers who own one pledge their allegiance.

White Horse also makes this excellent and very popular hydraulic gang plow.

Below: Eric and Anne Nordell have customized their implements to suit their market garden. Here is their forecart with a trail model single bottom subsoiler.

harrows

Pioneer Equipment of Ohio has been making outstanding spike and springtooth harrows for many years. I own these units and can't say enough good about them. They are exceedingly well built.

An old reversible orchard disc.

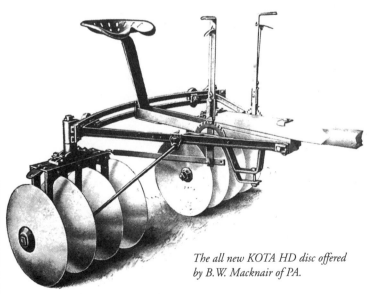

The all new KOTA HD disc offered by B.W. Macknair of PA.

Left: A Macknair display at HP Days featuring one of their KOTA discs.

Groffdale Machine Co. of Leola, PA is building all new discs and rebuilding and retro-fitting old-style field discs.

The unit above and left is their M series transport disc.

The small photo straight top is one of their KBA John Deere discs.

harrow carts

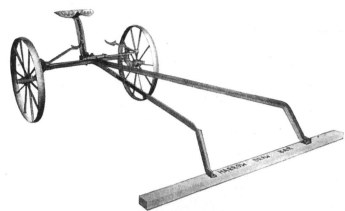

One old idea which could use some modern touches is the harrow cart. As you can see from the photo below of Tom Odegaard, being able to ride behind the harrow, instead of that precarious perch of standing on a tied down board, is the cat's meow. These old carts were all articulated so that the wheels "followed" around the corners. It would be a realtively easy matter for one or more of our new HD companies to build new versions.

Photo by Fuller Sheldon

A grape hoe and attachments from Australia. Neat idea which deserves a fresh look.

ideas for new implements?

A old one horse clod crusher.

Roller packers could be made new for us who prize their work.

cultivators

Eric Nordell's cultivator at work.

Paul Birdsall of Maine owns this team and cultivator.

I & J Manufacuring of PA. makes an all new lineup of culitvators ranging from one horse walk behind to one horse riding to straddle row to multi row. Their tools are well used throughout Amish communities and are earning an outstanding reputation.

The combination row and field cultivator on the left is just one of several being made by farm shops in Canada and the U.S.

drills and planters

The Pequea Company, see resource directory, makes an excellent line of large seed planters.

Old style drills and planters still abound. But manufacturers are hard at work on updates.

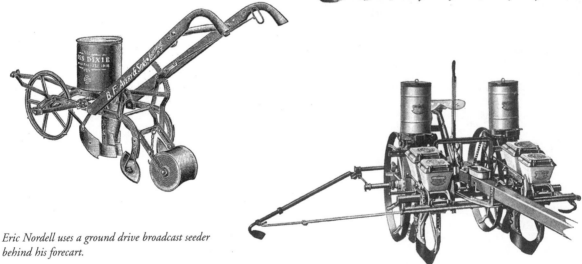

Eric Nordell uses a ground drive broadcast seeder behind his forecart.

manure spreaders

New manure spreaders are being made, and by several excellent firms! Whether your interest is in a completely refurbished New Idea wooden box, or a poly box sealed-bearing unit or a great big PTO steel spreader for forecart use, they are all available.

The Knob View brand spreaders on the left and below both feature a non-corrosive poly box and Timpken bearings.

row crop tools

Hogback Produce is one new young company working on some exciting new horsedrawn innovations. The tool on this page creates a raised bed and lays down plastic mulch all in one operation.

With the burgeoning promise of market gardening, a tool such as this, and the one on the next page, would add tremendous effeciency to the horsepowered operation.

Hogback Produce Company, see resource directory this volume, has also come up with this amazing horsedrawn transplanter. Warm water fills the holes which are punched in the plastic mulch. Two people ride on the back and gently press in the plants from trays on the slide.

Eric Nordell's unique roller, set up on a cultivator frame, does a great job of illustrating that the innovations of the future are going to come, mostly, right off successful farms.

mowers

A conventional Internaional #9 mower follows a haybine at HP Days in PA.

This #9 mower has been "souped up." Those steel wheels, mounted to the rubber tire spiders, are oversized to add more speed to the cutter bar.

Slick new after market parts are being made availabe for upgrading the old mowers.

Motorized rotary mowers rigged to pull behind forecarts...

... plus reel-type lawn mowers mounted in gangs ...

...all redefining what can be done with lawns...

...and perhaps even golf courses!

rakes

Of course the old stand by side delivery rakes, PTO and ground drive, still abound. But again the motorized forecarts make the modern spider and finger rakes attractive to some.

This ground drive tedder is very popular in the midwest.

Fjord horses on a motorized forecart handle a PTO tedder rake.

balers

At the first Mt. Hope Horse Progress Days, the ground drive baler above was demonstrated. This was a standard John Deere baler with a drive wheel conversion which worked, even on a slight hillside, exceptionally well. I rode on the baler and got this picture of the four Belgian horses stepping off and up to start the baler. Within a step and a half it was at full force! Impressive, but just the beginning.

Motorized PTO forecarts are making all tractor implements, though expensive, at least theoretically approachable.

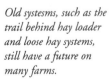

Old systesms, such as the trail behind hay loader and loose hay systems, still have a future on many farms.

bale movers

Round bales are unwieldy and some thought, a few years ago, that it would be difficult to figure out a way for horses to handle them. Not so. There are 7 or 8 variations on the theme of round bale mover. On this page we show just three. Photos by Lynn Miller

logging equipment

Many new types and styles of logging carts and arches continue to be developed and marketed.

misc.

White Horse Machine's subsoiler and cart point to the "no limits" future for horsedrawn implements.

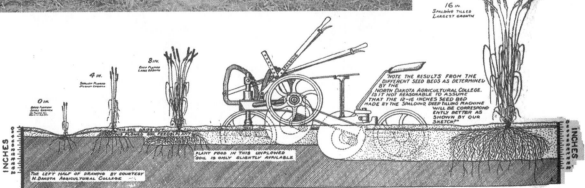

The Spalding literature had many nifty illustrations to dramatically portray the net effect of its operations.

Yes, there has been a huge increase in the availability of new horsedrawn equipment. But that does not mean a death to the market for the old standby implements. Don't forget that during the first third of the twentieth century millions upon millions of plows, mowers, rakes etc. were manufactured. There are hundreds of thousands of these tools hidden away in barns and still more littering the fence rows of North America. Auctions remain a vibrant and vital subcultural marketplace for usable implements. So long as horsemen gather, as below, and compare notes and wishes and wants the old and the new will keep on. The future is bright and growing.

Photo by Wendi Ross

Chapter Seventeen
Work Horse Diary
& Closing Thoughts

Working horses day after day, the routine wants to remain constant but it varies, and it can be different and yet similar in important ways. Here's a narrative, as personal diary, with added observations.

The sun's not yet up. The sky is a cold silver-tinged with reds. Even before I step outside, I can see from the kitchen window that all eight horses have their eyes glued to the door of the house in anticipation. They are looking for me, for my approach, for my errands of service to them. As I walk from the house, towards the shed which serves as our barn, the horses begin to nicker from their night pens. I slip from a low down deep throated hum to a soft whistle and back to a hum again, this morning it's an Argentinian Tango. 'Lucky', the Australian shepherd, follows me, excited for responsibility. The horses bob their noses and shuffle front feet as if to say *'it's about time'*.

Inside the barn/shed I pitchfork large quantities of the loose hay into each double manger. The two-foot deep mangers run across the front of each double tie stall which measures ten feet both directions. Each double tie stall is partitioned from the next by a planked half wall. The floor is packed sand. For fifteen years I have promised myself a real barn with a wood floor and well placed harness storage crowned by a large hay loft. But finances and circumstances haven't allowed such an improvement, so the horses and I make do. No, excuse me, it's better than that. We are thankful because what some might see as a crude open equipment shed temporarily converted to stabling for work horses we have enjoyed as an airy and well lit little barn. It has served us economically and well. I lean on the pitchfork and look around wondering how I will remember this building...

Horse's nicker louder and I wake up to the job at hand. First I go outside with halters and lead ropes to the pen which holds the two geldings. They are in a rail enclosure just 100 feet from the barn. It's their night accommodation. In the pen each gelding stands close by, nose forward as if aiming at a coming halter. I smile thinking about how these daily routines work so well to build the best working partnerships with the horses. Halters on and off everyday. Horses become not only accustomed but, by association, actually pleased with the process. When it comes to the haltering they each know that it is the first step in going to their stall and eating, so they lower their heads and reach their noses forward for the positively associated halter. And when day is done and halter is to come off once again, they lower their heads and now rack or rotate sideways, ears over towards me and down, to make it easier for the crown strap to come off, anticipating the comfort. They are happy going in to the barn and they are happy going out. An excellent indicator of a system which works.

Both horses haltered and with lead ropes in hand we exit the pen. They are anxious but they know, because I never allow exception to the rule, that they must walk behind me and on a loose lead. No bolting, no dragging back, no stepping on me. We walk together easy and fluid. It's either that or I get cranky and dream up clever ways of making their rule violations uncomfortable for them. They elected me leader. It may have been a rigged election, but I am the leader nonetheless. If I fall out of favor and dominion over them, it is my own fault and I must work to regain the leadership position. Otherwise, it is my contention, there can be no safe working relationship.

So we walk calmly to the barn where I snap each one to their individual secured manger chain which is bolted beside the 2 x 4 grain box. (These boxes are anchored in at the left and right sides - or opposite sides - of the manger, one each for each horse.) There is enough slack in each manger chain so that the horses might reach the middle of the manger and not much further.

The chains have big heavy bull snaps. I don't want them to break. It's not because I'm afraid my horses will try to break them but because accidents may happen and I prefer not to have loose horses in the barn. I occasionally use ropes instead of chain and sometimes *panic snaps* which allow that a twist or quick downward jerking motion will release a tangled endangered animal. These I usually relegate to trainees or new horses I am unfamiliar with. Just a small dose of caution which may one day pay off in a big dividend by saving a horse's life. But to date, after 30 plus years,

I have not had to use one in an emergency.

Geldings are snapped in and eating the morning hay. I flip off the electric fence charger and head back out to fetch a team of mares. With the girls you never know how the day's chemistry might stack up, one day mellow, the next day sullen or cranky. Today, with the first two, it seems they are mellow.

There are two pens wired with electric fence tape. Because I have four teams of horses in for haying, and because one of the mares beats up any horse she has to share space with, I needed additional enclosures and have set up a temporary electric pen arrangement for harmony and my own convenience.

This season, in total, I am working five of my mares and three geldings. Four of the mares work in teams or as three or four. The fifth mare has issues, not with me or any human but with other horses. Before I bought her, she was made a pet of and spoiled terribly. What she wanted she got. She works well with her gelding teammate as long as I am around to warn her off her worst tricks. She listens to me. I make sure of that. I've had her a short time and am determined to trick her into changing her attitude and nature. Right now she gets a pen to herself and occasionally is put into a large box stall for isolation. What I'd like to find is that she develops an attraction for one of my other horses and wants to be with it at all times. When this happens, I will use it to advantage.

We are only in our second week of concentrated work. Not until each horse is putting in a full 8 hour work day will all the minor anxieties level off. But one thing shouldn't change, each morning every horse should be anxious about getting to the feed. If one shows no interest, this could be an important first sign that something is not right. It might be protracted fatigue, it might be sore muscles, it might be any number of physical ailments. Best time to deal with potential health problems is at the first indication.

But this morning everyone is alert and salivating in anticipation of the clover/grass mix hay and the grain they expect is coming. I intentionally choose the pen with the two younger mares because I want to avoid yesterday's little incident.

I had gone after the older mares first, haltering and opening their wire pen gate and leading them towards the stabling when to my surprise and disappointment the younger ones pushed through their wire enclosure. It was as if nervously watching me open the other pen they figured out that the electricity was off. I shook that thought off as ridiculous, how could any horse come to such a conclusion? The thought nagged at me though, so later yesterday I tried a series of little experiments and determined conclusively that one of the two younger mares would watch and if she saw me handle the electric fence wire tape in any way she would then lean hesitantly against the wire herself. If she felt no shock she'd lean into it until it snapped! I have trained this young mare from a yearling and have always marveled at her intelligence, so my surprise was somewhat muted. I made the decision to outsmart her by always either leaving the charger on or removing her first and putting her back in last.

Both haltered, I toss the wire gate out of the way and head through the opening. One of mares pushed forward to get past me and I gently swat her chest with the loose end of the rope, she backs up. The smart mare makes her move, head down, for a alfalfa shoot at the lane's edge. She catches me off guard and I get drug half a foot. "What are you doing?!" I grunt as I jerk her lead rope and finally regain the composure I like to brag about. When we get to the big doorway of the open barn they of course tense up in anticipation and I use the moment to remind them of our election results. "I'm the leader you follow, remember!" Turning to face them, while switching leads in my hands, I stop them at the door.

They can now see the geldings happily munching away in their stalls. Facing them down and using quick matter-of-fact little jerks on the lead ropes I repeat the "Whoa" command until, when they come to fully recall the routine, they relax completely to accept that we aren't going inside 'til I say so. When I see the heads drop slightly and the ears go limp, I lead them in. This little exercise, I have found, whether on the lead or when ground driving back into the barn after work, pays huge dividends to reinforce leadership and frequently results in horses who will stand quietly for me in **any** circumstance.

They in their places in the stall, I go back outside to get two more mares. These two remain wary of the fence, on or off. And they've spent fifteen years with me so the routines are well understood. We make a quiet uneventful little walk to the stalls except for a wave of emotion I feel to think that one day these two very fine horses will, as others have, be too old to work with me. I will miss them and the assurance they always give me.

Last I retrieve the separated mare and gelding, this morning tying them side by side in a double tie stall. I will be working in the barn and know that a repeated

word of warning will forestall the angry mare from pestering her teammate. If I were to be leaving the barn for an extended period, I would choose to tie them in separate stalls.

Up to this point none of the horses have received any grain. This is intentional. I want them to have chewed forage in their digestive tract before they start to eat their grain. In this way the grain is slowed in its passage through the equine stomach allowing for better digestion. Long ago, a visiting veterinarian pointed out the whole oats apparent in the stabled horse manure. He suggested this timing routine (hay before grain) and a preference for rolled grains over whole. I have followed this suggestion for over thirty years with good results.

So this morning I go to the locked grain room and remove the lid from the galvanized garbage can I use to store the COB (corn, oats and barley with molasses.) At the first slight sound of the lid lifting a chorus of pleading nickers are aimed at me. I never tire of the sound, that deep soft edged percussive staccato hum, simultaneous from several horses. It comes as a full acknowledgment of my place in the relationship. They seem to say "Yes, you, please. I'm ready, can I have my grain now, first, before the others? I'm over here, where do you want me to stand? Please don't forget me. I need you to remember me, I need you to bring that grain to me. Bring the grain now and we'll see how the rest of the working day shapes up."

I fill a bucket with the grain and go from stall to stall, grain box to grain box, measuring out amounts I deem appropriate for each individual horse. Inside each wooden grain box there is a small salt block. I slide it to center of the box and pour the grain over the top of it. I have this unverifiable suspicion that this simple act slows the horses from quickly anxiously scarfing down the grain, spilling some on the stall floor. Watching their nose, lips, and tongue navigate the grain piled around the salt block, it seems like the rhythm and speed of their eating are natural.

Then I pause to listen to them, munching, snorting, breathing, pushing the salt blocks around the box. The view of their lovely forms, comfortable in their stalls and with their routines, is a view I never tire of.

Next I visit each horse with curry comb and brush. Had there been any concerns about sore shoulders or harness rubbings from the day before, I would take this time to check on their status. But this morning all is okay. As they eat I brush down each one in preparation for harnessing. Consumed as they are with eating I don't expect them to pay much attention to the naturally pleasant sensations of the morning brushing. If I should notice that one of them stops eating and acts as though particularly intense relief or discomfort is directly associated with a certain spot on their body I am going over, I pay attention, especially if it is anywhere that the collar makes contact.

My goal is always to get the horses fit and keep them in the field working. That means paying close attention to their comfort and care.

Next the harness: With no exceptions, each regular member of the working lineup has his or her own harness and collar. Today all the horses in the barn are mature and have worked long enough that I do not expect significant changes, day-to-day, in their collar fit. Sometimes, fat horses early in the working season will go down one to three collar sizes (inches) within a month to month and a half of hard work. Their necks carry a significant percentage of excess weight and as they sweat and work off that weight, the neck becomes thinner and shorter in depth.

There are many aspects of the harness horse system which are important, even critical. The fit of the collar rises to the top of that list. I pay more attention at the beginning of the season, but even now I frequently check to see how my collars are fitting. I want a perfect fit. If the collar is too tight, it will choke the horse down as he pulls and he'll quit from lack of breath. If, on the other hand, the collar is too loose (by a little or by a lot) the horse may continue working and you won't discover a problem until a sore has formed and the horse is in pain when it pulls. The damage from ill-fitting collars comes, 90% of the time, from the collar that is too loose. And it can be tricky because, sometimes, a collar may look and feel on the standing horse to fit perfectly, YET when that horse pushes forward it gets either too big or too tight. When I find a collar that fits my horse I make sure I remember where I put it because it's the first one I want to return to when I harness up.

I've decided that this morning I will be working the two teams of mares first. The plan is to take the young ones out and open a hay land (in our case a strip 150' wide by a quarter mile long). I'll be going through a low area where there might be some standing water and tough mowing. These girls, I know, will keep the pace exactly where I want it, when I want it. The evening before, on the better mower, I sharpened the sickle, hit the grease zercs and filled the oil jug. It's ready. But I'm off the subject, back to the job at hand, harnessing...

In the tack room I get down their two collars, each with sweat pads fastened in. I run the flat of my hand over the inside of each sweat pad feeling for anything sharp or aggravating. I take one collar in the stall along the left side of the left horse. Leaving the collar fastened, I unsnap the mare's stall chain and slip the collar, right side up, past her head and down over her neck into place, careful to pull mane hair out of the way of the top seat of the collar. Then I fasten her chain back up. During this whole procedure this mare stands quietly and accepting, pausing from eating just long enough for me to do my job. I have known horses who would never put up with having the collar go over the head. I didn't raise and train them. Someone else did. And somewhere along the line they decided, out of fear or obstinance, that no one was ever going to put a collar over their head. In those cases, I unbuckle the collar and push it up from the underside of the neck and refasten it topside, a procedure that is perhaps the safest bet for beginners but adds a half minute to my chores. And I like streamlining the process as much as possible. (I chuckle to myself realizing again that I seldom follow the rules and guidelines I have long given out to students at my workshops.)

Next mare gets her collar on same way, nice and quiet, over the head.

I return to the tack room where the harness is hanging on two big spikes driven head high in the wall. I pull down the brichen assembly from the one spike, put it up on my right shoulder and run my right arm under it and down the underside middle of the harness till I grab low the right side hame in that same hand. Left hand takes the left hame about in the middle.

(All the miscellaneous straps, lines, bridle etc. have been attached, hung, tied, or fastened in such a way as to make my carrying the harness as uncomplicated as possible. And I do it the same way each time I remove a harness. It's very easy for me to tell when someone else has been dealing with my harness because things are out of place. When everything is in the place I want, this harnessing process goes quick and easy.)

I carry the harness out and approach the stabled mare from behind. "Get over honey," is my command. She should step to the right and up near her teammate instantly, but she doesn't this morning. "Get over!" I say with more emphasis, and she complies. Walking up on her left side, I lift the right hame high, pointing its bottom skyward, while pushing it and the harness up on her back. The hames go forward to seat in the rib-lined groove of the collar, with the connecting top

North Dakota plowing scene. Photo by Fuller Sheldon.

hame strap at center top of the collar. Backing away slightly, I push each section of the harness, from off my shoulder and arm, up onto the waiting horse's back. Now it sits in a somewhat organized tangle atop the mare. Moving forward in the stall I ask her to back up and I pull the two hames to their seated positions on the collar. I'm careful at this point to see if I have accidently put a line or harness piece in under the hames. And I am also looking to make sure that the hames are properly positioned. They need to be equal on both sides, with the tug clip centered over the reinforced draft point of the collar. Everything is right, so I thread and tighten and buckle on the bottom hame strap.

The breast strap/pole strap assembly I prefer is removable, which means it fastens on both ends via heavy snaps. It's hanging fastened on the left side, so that when I throw the harness over the waiting horse, there is one less piece to concern myself with. At this point in the process I snap the right side of the breast strap to the bottom hame ring of the right hame. I leave the pole strap to hang for a second while I go back pulling and straightening the harness, gathering the brichen back over and under the tail. I check to see if the belly band is hanging straight down from the tug on the right side. Now I go forward, gather the pole strap between the front legs while reaching under for the belly band. The belly band goes over the pole strap and buckles in loose. Hanging from the two ends of the brichen are adjustable quarter straps with snaps on the ends. I fasten these to the ring at the end of the pole

strap. My horse is harnessed.

I repeat the process for the second mare. To read back over the process description, it seems complicated, however, I have repeatedly timed myself and when all things are as they should be, it is simple and harnessing one horse takes between one and two minutes.

With the next, older pair of mares, there is a slight deviation in routine. One mare's neck is relatively small for her bulk, and her head is quite large, with lots of width at the eyes. This means that, though she's perfectly willing to let me try, it is close to impossible to put the collar on over the head (whether rightside up or upside down). So for her, I unbuckle the collar at the top, unclip the sweat pad from one side and pull it out of the way. I then pass the collar up at the neck, bringing the pad over and into place before buckling the collar together. The remainder of the harnessing routine remains the same as with the previous team.

The third team takes a little longer this morning because I need to find a better fitting collar for the one horse. I didn't like the way it rolled up on his neck yesterday. So I try a couple of collars on him that have a different shape, though they are the same size. One has been stretched out wider and the other is a full-sweeney style specifically designed for a thick neck. I find the right collar, a good older one, with a thicker overall construction and it requires that I lengthen out the top hame strap. That done, I proceed with the harnessing until all eight head are outfitted.

Next I separate the difficult mare to a single stall before I take out the young mare team. I don't want to come back later to the barn and find her gelding teammate bunged up and something broken, all because she slipped once again into her *'get away from me!'* attitude.

Though there is a time and place for a single stall, I like the open double tie stall. It gives me a great deal of convenience. For example, I can drive a harnessed team directly into the stall when returning from work. And I can bridle, rig the lines, set the check reins, and back the team out of the stall when it's time to go to work.

Whenever possible, I work my horses without halters. Many teamsters prefer to leave the halters on under the bridles. I think this must add some discomfort to the horse and so I take them off whenever feasible. Mares ready, we back up and swing into the barn lane and walk out towards the waiting mower.

I walk the right mare over the mower tongue and swing the pair into place. They are standing now, either side of the tongue, exactly where they need to be for hitching. I walk alongside the left mare and spread the two lines across her back where they will be easily reached if needed. Then I proceed to the front of the team and raise up the tongue and neckyoke (the neckyoke is secured to the end of the mower tongue). I snap the breaststrap/pole strap assemblies to the neck yoke and take one quick look over the horse's heads to see if everything is okay. Next I walk back around the left side, picking up the driving lines and, with them in hand, proceed to hook the trace chains to the single trees. All this while the horses are standing calm, quiet and attentive. (The 'attentive' is important because I have found the inattentive horse is the one likeliest to jump when surprised, shocked or spooked by some unexpected sound or occurrence.)

All hitched, I climb aboard the mower, gather up the slack in my lines and check the team's ears. I want them both listening to me. Usually, the little vibrations, as I gather the lines, will tell them something's coming or that I'm getting set to ask them to go. Feeling my preparations, through the lines, can on occasion cause the horses to second guess me, and make them want to start before I'm ready. Seems natural, good intelligent working partners sense you are ready to set out, why wait for a formal command, why not step right out? Nope, don't ever let that happen. They will remember it and take charge and that's bad news, that's the beginning of unraveled. It is my contention that we train our horses every moment we work with them. If we forfeit the opportunity to say, and thereby control, exactly when our horses step

forward, we 'train' them to go whenever 'they' wish. A dangerous precedent. And one so easily avoided. It starts by **always** insisting *'we go when I say so and not before'*.

My young mares stand quiet, attentive, assured, ready. I smile and feel my breath shorten deep in my chest, it's not apprehension or fear or any negative reaction, it's that tightening that comes as a prelude to the waves of unavoidable natural gratefulness. I am so fortunate in my partnership with these beautiful creatures.

I gaze around til my focus returns. I give the command, *"okay ladies, let's do some mowing,"* chuckling to remember those hundreds of times I have admonished my workshop students to keep the verbal commands to their horses simple. And to always use the same sound or word for each desired action. *Don't do as I do, do as I say*. It's simple business and it's a complex craft.

We walk out to the hayfield and to that spot where, yesterday, I had tied a flag to the fence on my side. Also yesterday, I had paced off a new hay land marker across the field a quarter mile away, and propped a long stick in the cross fence. From the flag marker I am able to see across the relatively flat field to the stick. I swing the team in place with an effort to have my back be lined up, center, with the flag marker.

Sighting down the mower tongue between my horses, I look over the 1/4 mile wide hayfield to the fenceline on the other side and line the marker stick up to a third point on the horizon, a tall tree. From experience I know that getting a straight first cut will depend on my keeping the distant stick and the far distant tree lined up and in my sight at all times. If I simply aim at the distant stick, my cut will wander.

Points spotted, I do a quick look over the team and lines, lower the cutter bar from the carrier rod and squirt oil over its length from the oil jug. Next I climb on the seat, lever the bar down to mowing position and kick back the pedal to put the mower in gear. Checking to see if ears are back, I speak to the team and we head out mowing. Smooth, quiet, certain. Everything is as it should be.

It's not always this way. I know first hand that mowing can be a frightening procedure for man and beast, especially if either or both are unaccustomed. I've been doing it for over thirty years now and this particular team has been at it for five years. We make it look simple and safe. And that is how it can be. But I always worry that first-timers will get the wrong idea from such a relaxed picture and jump in to certain hazard or disaster. The picture must be earned.

We're mowing at a brisk walk, about 3 miles an hour, which is my preference. Even with these McD high gear #9 mowers timed, tuned and sharpened properly, going too slow can cause plugging in certain fine, wet, and/or wiry grasses. This is also why, when I'm opening a new land, possibly with wet lodged low spots, I prefer a team that will respond to my commands to walk faster AND be willing and able to stand quietly for a long stretch if I should need to clear a plugged cutter bar, or do some field mechanicing.

The team, on this opening pass, is walking through standing hay, belly deep. They love this run because frequently there is grazing available at a comfortable nose height. They've learned, over time, that if they keep a steady no nonsense straight ahead pace they can steal mouthfuls of grass and legumes as we mow. Fact is, I happen to know that they honestly think they are outsmarting me. They think I can't see. They think I don't know they're stealing bites. If they should stop or slow down to grab a bite, I would scold them. So they move along perfectly, trying not to let on that they are beautiful, clever thieves. As easy as they are on the lines, and with the snaffle bits, chewing and walking is no stretch for them. Our syncopation is built in small part on comic tolerance.

At 3 miles per hour we cross the field in 5 minutes. I never look back while mowing, keeping my eyes fixed on those two distant points until halfway when I've picked out a third point midway between the others. As I get closer to the end, I leave off looking at the stick in the fence and just concentrate on the mid-point and the far distant tree. Not until we reach the fence do I stop and allow myself to look back.

Wooee, is that pretty! Straight as an arrow, mown hay laying back in a combed and symmetrical pattern. Feels mighty fine. I don't call this working, I call this making art.

I turn left, cutter bar towards the fenceline, and mow the 150' wide end and lift the bar, still in gear. We step straight ahead a short distance across the previous land, where yesterday's hay is waiting for the morning dew to pass (I hope to rake this hay in the afternoon). When the new mown hay has vibrated off the cutter bar and while we're still moving ahead, I kick the pedal forward and take the mower out of gear. We swing around on a U-turn to head back from where we came.

That first cut opened a new long hay land, a strip of hay up against previous strips, all part of a forty acre field. Each land is approximately 4 acres. I work this way deliberately rather than dropping the entire forty acres at one time. This allows me to mow four acres, next to another four acre land that I am raking which is adjacent to yet another land where the hay is being picked up or buck raked. I get better quality hay this way and fret a whole lot less about losing the whole field to weather or other uncertainties. If anything gets damaged or lost, it's usually just four acres. It also works very well with the horses as motive power. In fact work horses brought me to these sorts of conclusions, led me to thinking of patterns of working that give me the best chance of comfort and success.

Same trajectory, opposite direction, when we get to the land edge I aim the team to walk over hay they just mowed. I'm using the foot pedal to hold up the cutter bar. We stop just before the cut edge. I lower the bar and put the mower in gear. I speak to the team and we head off across the short end of the land. Twenty feet in I can see a ball up on the bar and that we aren't making a full cut. I tell the team to stop, take the mower out of gear, and use the lever to raise up the cutter bar. I can see what looks like a big nest plugged on the ends of two center rock guards. Off the mower and with team lines in hand, I clear the ball of nest and hay off the bar and run my fingers over the tops of the guards. I feel and hear something. Going back over with my fingers, I find a loose guard. I tie a half hitch of the lines onto the lifting lever and retrieve my crescent wrench from the tool box. After the guard gets tightened, I oil up the cutter bar again. Usually don't do this until I've made a few rounds, but I'm down and it's handy right now.

Mower fixed, I take the lines in hand and, speaking to them, walk around the left side to the heads of the mares. I offer them each a handful of the new mown hay as I spot check their bridles and harness from the front. I look up and around to make sure no one is looking and I plant a kiss on each soft nose. If anyone did happen to see me do it, I would deny it ever happened. Gives the wrong impression, to other people that is.

Back on the mower, we cut the remainder of the land end and I make a clean corner and head up the long side. The opening pass, having been made in the opposite direction, has lain the hay down in such a way that I can expect, on this cut, a couple of ballups and sure enough, one comes straight away. The inside heal has gathered a knot of hay and we're missing a strip. I stop the mower and this time I back up the mares just a foot before I raise the cutter bar. Usually this will clear the knot without my having to get off the mower. It works and we set out again.

We mow for two hours and drop more than half of the land before we head back towards the barn. I have more horses than I need, and some of them are out of shape. They need time in the field. This team could keep going all day and drop up to ten acres of hay if I pushed them. But no need to. I'm gonna go back and get another team to mow with.

I drive the mower over by the barn and point the team away from the barn door. I'm careful to park in a spot where the mower won't roll either direction when I unhook. The easiest way to tell this is to stop and back up just a hair to see if, with the tugs slack, the mower remains put. I get off the seat and unhook one tug when I notice someone driving up the driveway. I hook the tug back up and wait til the visitor gets out of his pickup.

"Cool! Those are Clydesdales aren't they?"

My Belgian mares roll their heads and shrug their shoulders both letting out a deep sigh. The visitor wants to help me with the horses and I tell him firmly that he can talk to me all he wants but I must insist he not stand in front of the horses or touch anything while I unhook. We visit while I unhook the traces from the evener and move, lines in hand to the left side. I drop the lines on the ground and go round to unhook the heavy tongue from the breast straps. I'm careful to hold up the tongue until completely free and then let it down slow to avoid hitting their legs or hooves. I unhook the check reins and go back to my driving lines. We continue to talk. The mares think it's time to walk off and because I'm talking, they catch me off guard. They succeeded in walking three steps ahead, so I stop them and calmly start up again and walk them around, they think they are going in the barn, but I turn them and we walk over the tongue and I make them stand as though we are going to hook back up. I talk with the stranger for ten minutes until the one mare lets loose with a long squeaky methane blast that raises her tail in the air.

"Sorry girls, yes it is time to go to the barn." And we walk off quietly, with confidence and comfort and with the whole of our lives musically defined.

Thank you for reading my book. *Lynn Miller*

Work Horse Resource Directory

Please keep in mind that this list was compiled in the Fall of 2003 and is meant as a guide. Placement on this list does constitute an implied endorsement of product or service. And conversely, if a business or individual is not listed here it should not be taken as any negative statement.

BLACKSMITH

Anvil Magazine
PO Box 1810
Georgetown CA 96534
1-800-942-6845
www.anvilmag.com
anvil@anvilmag.com
Voice of the Farrier and Blacksmith

Centaur Forge Ltd
117 N Spring St
PO Box 340
Burlington WI 53105
(262) 763-9175

Green's Wheelwright & Blacksmith
David K Green
1211 17th Street
Penrose CO 81240
(719) 372-3226
Wheelwright, Blacksmith, Horse drawn
vehicles restored

BOOKS & VIDEOS

Anvil Magazine
PO Box 1810
Georgetown CA 96534
1-800-942-6845
www.anvilmag.com
anvil@anvilmag.com
Voice of the Farrier and Blacksmith

Diamond Farm Book Publishers
PO Box 537
Alexandria Bay NY 13607
1-800-481-1353
www.diamondfarm.com
Draft Horse, Donkey & Mule Books &
Videos

Engel's Coach Shop
Dave Engel
105 S Main
PO Box 247
Joliet MT   59041
(406) 962-3573
www.engelscoachshop.com
The Art of the Wheelwright on Video

Farmer Brown's Plow Shop
James & Bob Brown
10809 Davis Road
Hunt NY 14846
(585) 567-8158
www.farmerbrownsplowshop.com
Videos.

Mischka Farm
PO Box 224
Oregon WI 53575
1-877-647-2452
www.mischka.com
Draft Horse and Mule Books, Videos &
Calendars

Small Farmer's Journal
192 W. Barclay Drive
PO Box 1627
Sisters OR 97759-1627
1-800-876-2893
www.smallfarmersjournal.com
Publisher of the quarterly publication,
Small Farmer's Journal, books on training
and working draft horses and mules,
plowing, haying, and tillage tools. We also
have books on raising farm animals,
restoration and driving of carriages, sleighs,
blacksmithing, wheelwrighting, leather
working, gardening, growing produce, seed
sowing, soil, earth ponds, building and
renovating barns, and more.

COLLARS

Badger Brand Collar
RR 1 Box 86-2
Canton MN 55922
Farm & Buggy Collars, Adjustable Top
Collars, All Purpose Collars
Write for Catalog

Coblentz Collar Shop
3348 US Route 62
Millersburg OH 44654
(330) 893-3858 (answering service)
Collars

Shetler's Collar Shop
Harvey M Shetler
5819 Flat Iron Road
Conewango Valley NY 14726
Manufacturers of horse, pony & mule
collars, ornamental collars

Sugar Valley Collar Shop
18 Wagon Wheel Lane
Loganton PA 17747
Collars, Hames, Pads

EQUIPMENT

Balster's Implement and Parts Co
118 Third Street
Scotch Grove IA   52331
(319) 465-4141
Hard to Find Parts for: binders, combines, mowers, elevators, discs, plows, wagons, canvases, cutting parts, conventional plow shares, thresher teeth, disc spools, wood bushings, plow handles, Planter parts, singletrees, neckyokes, eveners, wagon hubs , wagon rear spindles, D. C. wheels

Beiler's Machinery
Elias & B John Beiler
601 Musser School Road
Leola PA   17540
(717) 656-9733
Dealing in new and used farm machinery, specializing in horse drawn equipment.

Boontown Sprayer
3384 TR 606
Fredericksburg OH   44627
Harry E. Miller, Manufacturer
Field Sprayer, Honey Wagon Spreader

Brookside Machine
CR 367 General Delivery
Berlin OH   44610
(330) 893-9212
Titus Slabaugh, Manufacturer
Manure Spreader

Leon M Brubaker
Rt 1 Box 693
Port Trevorton PA   17864
2 miles off 11 & 15 on Chapman Road
New & Used Horse Drawn Machinery, parts, plow shares, mower parts, Pioneer equipment

Conestoga Manufacturing
422 Mount Sidney Road
Witmer PA   17585
(717) 293-2716
Henry Esh, Manufacturer
Ground Drive Spreaders

Direnzio Equipment
8813 CR #3
Freedom NY   14065
(585) 567-4876
Manufacturer of logging carts

Easy Skid Log Carts
15235 Nash Road
Burton OH   44021
330-296-4244
Bill C. Fisher, Manufacturer
Logging Cart

E.M. Equipment
1964 TR 178
Baltic OH   43804
(330) 897-6006
Emanuel J. Yoder, Dealer
Mowers, Tedders

E-Z Spreader Manufacturing
1951 CR 70
Sugarcreek OH   44681
(330) 852-2666
Mose Erb, Jr., Manufacturer
Ground Drive Spreaders

E-Z Trail Manufacturing
7525 Harrison Road
Fredericksburg OH   44627
(330) 287-6992
Eli Hershberger, Manufacturer
Round Bale Carrier, Standard Forecart

Farmer Brown's Plow Shop
James & Bob Brown
10809 Davis Road
Hunt NY   14846
(585) 567-8158
www.farmerbrownsplowshop.com
For all your plowing needs: handles, cross bars, points, videos, walking and sulky plows. For all your logging needs: Farmer Brown's Multi-purpose Logging Arch, logging tools, Belgian horses, videos.

Forest Manufacturing
906 Daffodil Road
Reynoldsville PA   15851
(814) 894-5713
David E. Miller, Manufacturer
Logging Cart

Gateway Manufacturing
7656 E Colonville Road
Clare MI   48617
(989) 386-4198
Alvin Yoder, Jr., Manufacturer
PTO Power Cart

Gingerich Tractor
6128 SR 39
Millersburg OH   44654
(330) 674-0456
Mahlon Gingerich, Dealer
Tedders, Rakes

Graber's Country Store
18786 200th Street
Bloomfield IA   52537
(515) 830-2971 Voice Mail
Grabilt Grabber, Grabilt Stabber

Groffdale Machine
194 S Groffdale Road
Leola PA   17540
(717) 656-7657
Mowers, Disks, John Deere KBA Disks

Hoegger's Wagon & Harness Shoppe
Hoegger Supply Company
160 Providence Road
Fayetteville GA   30215
1-800-221-4628
www.hoeggergoatsupply.com
Wagons, carts, sleigh kits, harness

Hogback Produce
10221 Hogback Road
Fredericksburg OH   44627
John A. Miller, Manufacturer
Fertilizer spreader, disc, walking cultivator, mulch layer, produce sprayer, transplanter, sicklebar mower, trailer gear mower.

Horse Progress Day
Eli J C Yoder
445 S Mill St
Sugarcreek OH   44681
(330) 852-4603

Hostetler Manufacturing
6622 S Carr Road
Apple Creek OH   44606
(330) 698-1913
Aaron Hostetler, Manufacturer
Mulch Layers

Neil M. Hostetler
130 E CR 200 N
Arthur IL   61911
(217) 543-2217
100 HP Tiller

I & J Manufacturing
5302 Amish Road
Gap PA   17527
(717) 442-9451
Jacob Blank, Manufacturer
Cultivators, Slitter/Subsoiler, Plows, Harrows, Sicklebars, Tedders, Rakes, Forecarts

Imko Systems
2980 W Fenner Road
Troy OH   45373
(937) 339-5197
Monte Swank, Manufacturer
Horse Drawn Log Loader

Iowa Valley Carriage Supply
Sandy McKee
702 E Harrison
Toledo IA   52342
(641) 484-4784

Joe's Repair Shop
9108 Mount Hope Road
Apple Creek OH   44606
Dan Schlabach, Manufacturer
Ground Drive Spreaders, Plows

Kenda Custom Carts & Wagons
116 S 9th Street
Sunnyside WA   98944
(509) 839-6285
Manufacturers of all sizes of carts, wagons,
buckboards & carriages.

Knob View Spreader
4155 CR 59
Baltic OH   43804
(330) 897-1106
Roman Miller, Manufacturer
Ground Drive Spreader

Ron Mack
4399 Seville Road
Seville OH   44273
(330) 769-2960
Manufacturer
Ground Drive PTO Cart

B. W. Macknair & Son
3055 US Hwy 522 North
Lewistown PA   17044
(717) 543-5136
New and used parts for horsedrawn
equipment

Mascot Sharpening
434 Newport Road
Ronks PA   17572
(717) 656-6486
Omer S. Fisher, Manufacturer
Mowers

Midwest Leather & Harness Co
81202 Highway 70
PO Box 548
Beckwourth CA   96129
1-888-211-3047
Pioneer Equipment, Harness, Collars,
Halters, Harness Parts, Hames, Collar Pads,
Neckyokes, Eveners, Running W, Surcingle

Mill Machine
7684 SR 514
Big Prairie OH   44611
Ervin E. Yoder, Manufacturer
Logging Cart

Lloyd Miller
5438 TR 353
Millersburg OH   44654
(330) 674-4267
Farmer/Operator
Square Bale Stuffer/Wrapper, Discbine

M. E. Miller Tire Co
17386 State Hwy 2
Wauseon OH   43567-9486
(419) 335-7010 ext 4
Tires & tubes for hay mowers, manure
spreaders, hayrakes, corn planters &
pickers, garden tractors, older farm tractors
& implement wheels in stock

Miller Repair Shop
2945 S 050 W
LaGrange IN   46761
(260) 463-2352
Mervin Miller, Manufacturer
Manure Spreaders, Cultivators

Vernon Miller
10311 Trail Bottom Road NW
Dundee OH   44624
(330) 852-4691
Vernon Miller, Dealer
Shank Disc Chisel, Tedder, Round Baler

Mullet Repair Shop
7705 W 450 N
Shipshewana IN   46565
(260) 768-7935
Enos Mullet, Manufacturer
Rotary Plow Packer

Murchison Wagon Works
115 E Vista Dr
Silt CO   81652
(970) 876-2107
Wheels Built/Repaired, New Wagons,
Restoration, Hitch Wagons, Chuck
Wagons, Carts.

Newco Sales Inc
310 S Peach Avenue
Marshfield WI   54449
(715) 387-1015
newco@tznet.com
http://www.tznet.com/newco
Log Grapple Trailer Systems for Horses by
Majaco

Orenco Wagon Company
Rob Lewis
22930 NW Alder
Orenco OR   97124
(503) 648-3615
www.orencowagon.com
Wagon & coach manufacturer

Ox Bow Trade Co
Highway 395
PO Box 658
Canyon City OR   97820
(541) 575-2911
Vis-à-vis, Surreys, Wagons, Buggies,

Concord Stagecoach, Carts, Sleighs, Amish
Harness, Shaves, Poles, Parts

Pequea Machine Inc
200 Jalyn Dr
PO Box 399
New Holland PA   17557-0399
(717) 354-4343
www.pequeamachine.com
Manure Spreaders

Pequea Planter
561 White Horse Road
Gap PA   17527
(570) 442-4406
Gideon Stoltzfus, Manufacturer
Corn Planters

Pioneer Equipment Inc.
Wayne H Wengerd, Manufacturer
16875 Jericho Road
Dalton OH   44618
(330) 287-0386
Walking Plows, Sulky Plows, Spring and
Spike Tooth Harrows, Forecarts, Wagon
Gears, Steel Wheels, Wagon & Mower
Tongues, Steel & Wood Eveners,
Doubletrees & Neckyokes, Motorized
Power Carts

Pleasant Valley Inc.
1302 County Line Road
Venus PA   16364
(814) 354-6109
Monty Miller, Manufacturer
3 Point Hitch, Farm Logging Cart

RK Sales Inc.
8155 Ryan Road
Seville OH   44273
(330) 769-2731
Richard Stoll, Manufacturer
Gen-Til Aereater

Raber Equipment
5150 CR 229
Fredericksburg OH   44627
(330) 695-6793
Abe Raber, Dealer
Round Bale Wagon

Small Farmer's Journal Draft Horse &
Horsedrawn Equipment Auction & Swap
Meet
Usually the 3rd weekend in April – call for
exact dates: 1-800-876-2893
PO Box 1627
Sisters, OR   97759
www.smallfarmersjournal.com
auction@smallfarmersjournal.com
Draft Horses, mules, buggy horses, working

ponies, buggies, carriages, stagecoaches, sleighs, freight wagons, covered wagons, buckboards, pleasure driving vehicles, farm wagons, carts, fixer-upper wagons, restoration project carriages, carriage lamps, buggy & wagon parts including wheels. New and used collars, collar pads, halters, bits, bridles, harness parts, harness for all size animals, doubletrees, neckyokes, eveners, tongues, shafts, sleigh bells, blacksmithing and horseshoeing tools. Harness hardware, harness making tools including sewing machines. Horselogging gear. Horsedrawn farm implements of all sizes, types and conditions including: mowers, plows, harrow carts, discs, harrows, manure spreaders, seed drills, cultivators, stump pullers, etc.

Troyer's Windmill Sales
3981 CR 70
Sugarcreek OH 44681
(330) 893-0051
Mills, Towers, Parts, also Ornamentals

Victoria Leather Co.
217 S 7th St
Delavan WI 53115
1-800-990-8669
www.ruralheritage.com/sleighbells
Harness Hooks, Sleighbells

White Horse Machine
5566 Old Philadelphia Pike
Gap PA 17527
(717) 768-8313
Melvin and Henry King, Manufacturer
Plows, Subsoiler, Forecarts

Willeheim Acres
John & Carole Wylie & Son
RR 2
Vankleek Hill Ontario Canada K0B 1R0
(613) 678-3622
Forecarts, Logging Arches, Pony Forecarts

Woodhull Spreader Mfg.
4553 Old State Road
Woodhull NY 14898-9724
Manure Spreaders

Yard Hitch Inc.
RR 3 Box 112
LaCrescent MN 55947
(507) 895-8024
Gene Gunderson, Manufacturer
Hitch Cart, Grader Blade, Log Arch, Bale Spear

Yoder Equipment
19887 Jade Avenue

Bloomfield IA 52537
(641) 664-2797
Ivan Yoder, Manufacturer
Round Bale Grabber, spear

Jacob L. Yoder
5894 TR 606
Fredericksburg OH 44627
(330) 695-2340
Dealer
E-Z Trail Bale Basker

HAMES

John I Beiler Mfg & Sales
610 Lime Quarry Road
Gap PA 17527-9793
(717) 442-4627
Aluminum Hames

Sugar Valley Collar Shop
18 Wagon Wheel Lane
Loganton PA 17747
Collars, Hames, Pads

HARNESS

Bauman's Leather Shop
3518 Muddy Pond Road
Monterey TN 38574
(931) 445-3234

Big Sky Leatherworks
5243 Highway 312 E
Billings MT 59105
(406) 373-5937
www.bigskyleatherworks.com
Double trees, neckyokes, hames, buggy, work, logging harness, harness parts, bells, many driving supplies, bits, collars, pads

Bontrager Harness Shop
5913 E Greenfield Road
Haven KS 67543
(620) 465-2771 (answering service)

Joe H. Bowman & Sons
6928 County Road 77
Millersburg OH 44654-9181
Harness, Tack, Blankets

Bunker Hill Harness Shop
Samuel J J Schwartz
9100 South 150 East
Geneva IN 46740

Center Square Harness Shop
J. Samuel Esh
246 Forest Hill Road
Leola PA 17540
(717) 656-3381
Custom-Made Draft Horse Harness

F. H. Ebinger Co Inc
1 Peatfield St
Ipswich MA 01938
1-877-884-7200
fredco@ipswich.org
Leather Supplier to American Manufacturers and Craftsmen

Frantz Quality Nylon
Michael Frantz
3513 N SR 75
Camden IN 46917
Nylon Draft Harness

George's Harness & Saddlery
Box 29
Ryley Alberta Canada T0B 4A0
(780) 663-3611

Glick's Harness Shop
4211 Rugles Road
Fredericktown OH 43019
Write for catalog
Manufacturers of Quality Biothane Harness for farm, pulling, parade, driving, cart, show

Goose Creek Halter
3299 TR 406
Millersburg OH 44654
Beta lines, show or farm, team or single

Hog Branch Harness
Ken & Wanda Bauerle
PO Box 290
Purvis MS 39475
(601) 794-6149
www.hogbranchharness.com

Homestead Harness Shop
11085 Monroe Road 881
Paris MO 65275
(660) 327-4875
homestead@mcmsys.com
www.homesteadharness.com
Custom-fit leather harness, collars, hames, pads, halters, lead ropes, tongues, wagon bows, shaves, eveners & neckyokes, Pioneer Equipment

Jamesport Harness Supplies
Elmer L Beechy & Family
21776 State Highway 190
Jamesport MO 64648
(660) 684-6775
Harness makers, carriages & wagons

Kalona Harness Shop
David M Gingerich
824 6th Street
Kalona IA 52247-9486
New Harness, Nylon Halters

Maple Valley Harness Shop
2724 Co Hwy 31
Cherry Valley NY   13320
(607) 264-3138
New & used harness, tack & supplies, sales,
service & repairs in leather, nylon &
bioplastic, harness hooks, horse warmers &
supplies

Aaron Martin Harness Ltd
4445 Posey Line
RR 1
Wallenstein Ontario Canada N0B 2S0
(519) 698-2754
info@aaronmartin.com

Midwest Leather & Harness Co
81202 Highway 70
PO Box 548
Beckwourth CA   96129
1-888-211-3047
Pioneer Equipment, Harness, Collars,
Halters, Harness Parts, Hames, Collar Pads,
Neckyokes, Eveners, Running W, Surcingle

Meader Supply Corp
23 Meaderboro Road
Rochester NH   03867
1-800-446-7737
Harness, hardware, accessories, yokes,
eveners, halters, bits, bridles, blankets, leg
wraps, bells, health care & grooming aids,
books & videos, rubber mats

Miller's Nylon Harness Shop
10734 Shadeland Road
Springboro PA   16435
(814) 587-3226

Northwest Harness & Leather
Steve Henricks
HC 85 Box 94
Bonners Ferry ID   83805
1-866-267-8828
www.nwhl.net
Quality Farm & Carriage Harness
Leather & BioThane, Collars & Pads,
Accessories

Olson's Carriage & Harness
10855 Hodgen Road
Black Forest, CO 80908
(719) 495-4486

Peach Lane Harness Shop
Abner S Esh
Manufacturer of nylon & urethane plastic
harnesses
88 Peach Lane
Ronks PA   17572
(717) 687-5122 (answering service)

R Tiny Farm Harness
16509 Decker Creek Dr
Manor TX   78653
1-877-784-6932
www.rtinyfarm.com
Custom made Biothane & Beta Harness

Roy's Harness & Tack Shop
R. A. Royston
90 Muskoday Trail
Buckley MI   49620
1-866-755-6858
Leather & Nylon Harnesses

St. Paul Saddlery
953 W 7th St
St. Paul MN   55102
Harness, Saddlery Manufacturer since 1908

Samson Harness Shop Inc
6543 Akonerva Road, Dept S
Gilbert MN   55741
(218) 865-4602
Manufacturers of first quality show & work
harness, collars, pads, hames, eveners

Jacob K Schwartz & Sons
5364 Fisher Road
Conneautville PA   16406
Nylon Harness, Belgian Gelding Teams-
Singles

Shipshewana Harness & Supplies
815 N Van Buren St
PO Box 745
Shipshewana IN   46565
(260) 768-7254
www.shipshewanaharness.com
harness@shipshewanaharness.com
Blacksmith supplies, halters, blankets, lead
ropes, leather & BioPlastic Harness

Singletree Leather & Harness
Jim & Diane Dickinson
8284 East Hawley Road
Hesperia MI   49421
(231) 854-3845
www.singletreeproducts.com
singletr@triton.net
Work, Show, Parade, Driving Leather &
Biothane Harness from Draft to Mini
Dealer of Wells 5 Star Mule Pads, Beta:
Reins & Halter – Bridle Combos, Biothane
Breeching & Breast Collars, Mesh Haybags/
Nylon Tack Bags

Taborton Draft Supply
54 Taborton Road
Averill Park NY   12018
(518) 794-8287

Townline Harness Shop & Shoe Repair
John D Miller
12466 Townline Road
Windsor OH   44099
Driving, work, pony harness & parts

Victoria Leather Co.
217 S 7th St
Delavan WI   53115
1-800-990-8669
www.ruralheritage.com/sleighbells
Harness Hooks, Sleighbells

Village Harness
3578 W Newport Road
Ronks PA   17572

Warnercrest Farms Harness Shop
PO Box 92
Masonville NY   13804
(607) 265-3577

Wengerd's Harness Shop
6924 TR 323
Millersburg OH   44654
(330) 674-9323 (6 -7 pm EST)
Manufacturers of Buggy Harness, Regular
Draft Harness, Parade Harness & Show
Harness. All our Harnesses are made with
top quality leather and Biothane.

The Working Horse
Richard & Betty Schauer
16655 County 6
Park Rapids MN   56470
(218) 732-4208
Harness & Tack Shop

Yoder's Harness Supplies
Vernon Yoder
4819 Springdale-Hunters Road B
Springdale WA   99173
New and repair Harness, Saddles, Shoes

J. R. Yoder Nylon Works
3649 TR 159 – Route 2
Sugarcreek OH   44681
(330) 893-3479
Bio-Plastic Coated Nylon Harness, Show,
Parade, Pulling, Farm & Buggy Dirving
Harness

Yoder's Harness Shop
260 N County Road 575 East
Arcola IL   61910
(217) 268-3837

Yoder's Harness Shop
Levi Yoder
14698 Bundysburg Road
Middlefield OH   44062
440-682-1505

WHEELWRIGHT

Engel's Coach Shop
Dave Engel
105 S Main
PO Box 247
Joliet MT 59041
(406) 962-3573
www.engelscoachshop.com
The Art of the Wheelwright on Video

Green's Wheelwright & Blacksmith
David K Green
1211 17th Street
Penrose CO 81240
(719) 372-3226
Wheelwright, Blacksmith, Horsedrawn
vehicles restored

Greenman Carriage Company
Charles Greenman & Dwight Mitchell
Highway 218, 500 Monroe
PO Box 250
Floyd IA 50435
(641) 398-2299
wheelwright@mchsi.com
Wheel repairs, new wheels, restoration of
most horse drawn vehicles

WORKSHOPS

Chuck Baley's Four Corner's Driving
School. Teach people and horses how to
drive. Call or write for dates. (970) 731-
2431, PO Box 4357, Pagosa Springs, CO
81147

Doc Hammill's Workhorse Workshops.
Learn safety, non-confrontational techniques, and the basics of driving, while
using the horses and equipment of a
working horse-powered homestead. Contact
Doug Hammill DVM, (406) 756-2889,
PO Box 415, East Glacier Park, MT
59434.

Draft Horse Workshops at Fair Wind
Farms, Jay & Janet Bailey & Family. Our
comprehensive workshops will give you the
basic skills to own and work with draft
horses. (802) 254-9067, 511 Upper
Dummerston Road, Brattleboro VT 05301.
email: fairwind@sover.net

Farmer Brown's "Hands On" Basic Draft
Horse Clinics: Basic Draft Horse, Logging.
(585) 567-8158, 10809 Davis Road, Hunt
NY 14846.
www.farmerbrownsplowshop.com

Driving Drafts – Adult Weekend. Singles/
Doubles Beginner & Intermediate Levels.
www.kathysponies.com. Call (610) 838-
2928, Flint Hill Farm, Coopersburg PA
18036

Beginner's Horsemanship School, Eli J. C.
Yoder, 445 South Mill Street, Sugarcreek
OH 44681 (330) 852-4603

Olson Driving School. All levels welcome
to this hands on experience in driving
singles to multiples with carts, wagons and
farm implements. (719) 495-4486, 10855
Hodgen Road, Black Forest, CO 80908.
www.olsoncarriage@msn.com

Big Horse Ranch. Introductory to driving,
Learn to harness & drive. (904) 819-0243,
St. Augustine FL.
www.BigHorseRanch.com

Russells Workhorse Farm, Learning to Farm
with Draft Horses. Kenny & Renee Russell,
(601) 795-4200, 12055 Highway 11
North, Poplarville MS 39470.
www.russellsworkhorsefarm.com

Working Horse Breed Registries

American Brabant Association
2331A Oak Drive
Ijamsville, MD 21754
301-631-2222
brabant@ruralheritage.com
www.ruralheritage.com/brabant

Belgian Draft Horse Corporation of America
PO Box 335
Wabash, IN 46992-0335
(260) 563-3205
www.belgiancorp.com
belgian@belgiancorp.com

Clydesdale Breeders of the USA
17346 Kelley RD.
Pecatonica, IL 61063
Phone (815) 247-8780
Fax (815) 247-8337
clydesusa.com
secretary@clydesusa.com

American Cream Draft Horse Association
Nancy H Lively, Secretary
193 Crossover Road
Bennington, Vermont 05201
802-447-7612
Fax -447-0711
www.americancreamdraft.org
info@americancreamdraft.org

American Donkey & Mule Society
PO Box 1210
Lewisville TX 75067
(972) 219-0781
www.lovelongears.com
adms@juno.com

American Haflinger Registry
2746 State Route 44
Rootstown, OH 44272
Ph. 330-325-8116
FAX: 330-325-8178
haflingerhorse.com
Ahaflinger@aol.com

The Lippit Club. Inc. (Lippit Morgans)
P.O. Box 1921
Brattleboro VT 05302
802-257-4968
http://members.tripod.com/Lippit Club
needeep@together.net

American Mammoth Jackstock Registry
PO Box 1190
Enumclaw, Washington 98022
Phone/Fax 830-324-6834
www.amjr.us
register@amjr.us

American Shire Horse Association
Sheila Junkins, Secretary
PO Box 669
Glenwood Springs CO   81602
(970) 384-1511
secretary@shirehorse.org
www.shirehorse.org

American Suffolk Horse Association
Mary Margaret M. Read
4240 Goehring Rd.
Ledbetter, TX 78946-5004
(979) 249-5795
www.suffolkpunch.com

North American Spotted Draft Horse Association
PO Box 447
Chesaning, MI 48616
Phone: 517-204-5561
Fax: 989-845-2154
www.nasdha.net
spotted_drafter@yahoo.com

Norwegian Fjord Horse Registry
PO Box 685
1203 Appian Dr
Webster, NY 14580-0685
Phone: 585-872-4114
Fax: 585-787-0497
www.nfhr.com
registrar@nfhr.com

Percheron Horse Association of America
P.O. Box 141, 10330 Quaker Rd.
Fredericktown, OH 43019
Phone 740-694-3602
Fax 740-694-3604
www.geocities.com/phaoa/
percheron@percheronhorse.org

P.O.S.M. Horse Registry
PO Box 424
Machias ME   04654
All American Light Draft Horses

Measurements

general

HORSE-POWER: One horse-power is the equivalent of 33,000 foot pounds of work per minute.

One 1500 lb. horse at steady, regular work produces one "horse-power." Two 2000 lb. horses in good condition at steady work would produce three "horse-power." During a hard pull the same team would produce up to 30 horse-power.

A horse or mule at steady regular labor for ten hours per day can be expected to pull a dead-drag weight equivalent to 10 percent of the animal's body weight. A hard pull might equal body weight.

hitch gear

TONGUES: For full size horses, figure an average length of 9 ½' to 10' from doubletree hitch point to neck yoke point. Add length needed to fasten to implement or vehicles. Make sure that doubletree is sufficiently ahead of implement so that turning horses will not hit or tangle. Add 6" beyond neck yoke stop. Tongues should taper from at least 2 x 3" if wood.

NECK YOKES: The neck yoke is the same length as the doubletree for the same horses. Lengths vary from 38" to 48" for wagons (or desired wide-set). The narrowest would be from 28" to 30" for use in some plow hitches and other narrow setups. Neck yokes, if wood, should taper from the center out. The center should be about 3" thick with the ends at 1 ½" to 2".

SINGLETREES: For full-sized horses, singletrees vary in length from 26, 28, 30, 36 and 38 inches. The 36" and 38" lengths would only be used for very wide hitches, like wagons, if at all. 28" and 30" are the most common. The singletree normally measures 2 ½" thick at the center with ends at least 1 ½" thick.

DOUBLETREES: Wagon doubletrees are normally 42, 46 or 48 inches long. Narrow, plow-type doubletrees are anywhere from 30" to 40" in length with 28" rarely seen. A wooden doubletree should taper from the center and be at least 4" thick at center with something like 3" at the ends.

TRIPLETREES: 50" to 54" in length with hitch point one third over. 4" to 5" thick if wood.

CLEVIS: Are sized to fit.

SHAFTS: For full-sized horse – 28" in front, 5'6" to hold back hardware. 36" wide inside back. 100" long. See page 121 for illustration.

LEAD OR DRAW CHAIN OR CABLE: 10 foot adjustable to 12 foot.

FOR BUCKBACK SYSTEM – Tie chain = 60" to 72". Buckback strap: inside check 48"; outside check 54". Main strap 106" adjustable.

harness

Weight of farm harness (with 1 ½" tugs), approximately 75 lbs.

Weight of farm collars, 12 lbs. for 23 inch, 14 lbs. for 26 inch, 16 lbs. for 28 inch.

Tugs – vary in dimension from 1 ½" x 6', 1 ¾" x 6', 1 ½" x 6' 6", 2" x 6'6", up to 3" x 6'6".

Lines – vary from 1 1/8" x 20' to 1 ¼" x 20' for team lines.

Billets – average 1 ½" in width.

Pole straps – (sometimes called martingales) from 1 ½" x 4' to 1 ¾" x 4', some as wide as 2".

Belly bands are 1 ¼" to 1 ¾" wide

Hame straps – vary from 1" x 21" up to 1" x 36" or 1 1/8" x 24" up to 1 ¼" x 24" on up.

Breast straps are 1 ½" x 4 ½' or 1 ¾" x 4 ½' (doubled).

Bits for draft horses are usually 6" long, some 6 ½".

Trace chains are from 18" to 24" in length with 5/16" wire links. Some go to 36" with 11 links.

Hame balls are usually 2 ¼" to 2" in diameter with 1 1/8" opening.

Small Farmer's Journal

Established in 1976

featuring Practical Horsefarming

An international quarterly in support of small scale family farming,
organic agriculture, small towns, appropriate farming technology and
food justice. Supported 100% by readers. Please consider a subscription.

$35 per year in the U.S. purchases 4 issues of this vital and unique
Journal. (other countries add $20 post per year U.S. funds)

Small Farmer's Journal

P.O. Box 1627, Sisters, Oregon 97759
toll free 800-876-2893
www.smallfarmersjournal.com
fax 541-549-4403   ph 541-549-2064

Jess Ross on the buckrake with Robert Clark in the Big Hole region of Montana. Photo by Kristi Gilman-Miller.

TRAINING WORKHORSES / TRAINING TEAMSTERS

by Lynn R. Miller
is a text combining two books in one. This highly acclaimed text covers the subjects of: training horses to work in harness - on the farm, in the woods, and on the road, correcting behavior problems with work horses - and training people to drive and work horses. Features round pen training techniques. A companion volume to the Work Horse Handbook and part of the *Work Horse Library* project.
482 photographs and hundreds of drawings on 352 pages. Soft Cover $32.95.
Hard cover $50.00

HORSEDRAWN PLOWS AND PLOWING

by L. R. Miller. Already a classic! 368 pages with over 1,000 drawings and photos covering how to plow with horses using older equipment and new implements. Here you will find simple diagrams explaining tricky adjustments for both riding and walking plows. Detailed engineer's drawings of John Deere, Oliver, McCormick Deering, Parlin Orendorff, Avery, and many other older manufacturers will be immensely helpful to folks restoring equipment. Also includes closeup photos and information on new makes of animal-drawn plows including Pioneer and White Horse. Part of the **Work Horse Library** project.
$32.95 Soft Cover

(Shipping $5 1st book [$7 other countries] $1.50 each additional)
To order contact; SFJ, P.O. Box 1627, Sisters, OR 97759
800-876-2893   www.smallfarmersjournal.com   541-549-2064

HORSEDRAWN TILLAGE TOOLS: DISCS, HARROWS, ROLLERS, CULTIVATORS & MORE

By Lynn R. Miller

This important book covers operation, care and repair of animal-powered seedbed preparation and weed cultivation tools. Over 1,000 photos and illustrations covering the new and old, the useful and the historic. Includes important information on how animals are hitched and driven with these implements. There is tuneup and field operation information. A chapter on cultivation covers a wide variety of field and row crop options. 368 pages in all.

$32.95 Soft Cover
$50.00 Hard Cover

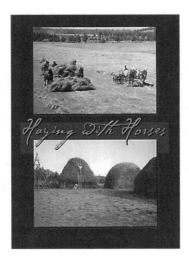

HAYING WITH HORSES,

by Lynn R. Miller.

The first comprehensive text on the subject. A new practical reference volume, covering all aspects of haymaking with horses and mules in harness. Offering in-depth information on Mowers, Rakes, Hayloaders, Buckrakes, Stackers, Tracks and Trollies for barns, Hay Fork systems, Balers, Wagons, Feed Sleds, and Forecart adaptations etc. And covering the building of loose hay stacks and wagon loads. Unloading systems, and feeding systems are also covered. 368 pages thick with 1,000 + illustrations

Soft Cover $32.95

(Shipping $5 1st book [$7 other countries] $1.50 each additional)
To order contact; SFJ, P.O. Box 1627, Sisters, OR 97759
800-876-2893   www.smallfarmersjournal.com   541-549-2064

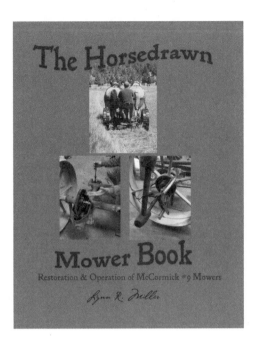

The Horsedrawn Mower Book

For horsefarmers and antique implement collectors, it is certain to be a most useful handbook. With hundreds of photos and drawings, the profusely illustrated text covers restoration, rebuilding, repair, and tuneup with a focus on the very popular McCormick Deering (International) No. 9. It also includes references to other makes and models as well as resource information for updating cutter bar assemblies to new materials and functions. Mr. Miller, along with being a long time horsefarmer, has restored mowers for 25 years and taught several workshops on the subject. *$32.95 Soft Cover $50.00 Hard Cover.*

FARMER PIRATES & DANCING COWS

A book of essays by Lynn R. Miller

"As with all great essayists, what is offered is the articulate half of a conversation. Our responses are anticipated, welcomed in advance....Farmer Pirates & Dancing Cows visits in the garb of a neighbor who knows he is among friends—informal, deeply engaged in the moment, garrulous, even opinionated, knowing it won't be held against him. His is the voice of the neighbor we all wish we had, driving hard at an important point that needs attention, yet wise enough at the end to give us the quiet for thought and resolve. Accepted and welcomed because we all need such a voice standing up for us, breaking trail out ahead of the heavy wagons to follow, full of the stuff of our lives."
- Paul Hunter, Seattle, WA poet

224 Pages, soft cover $15

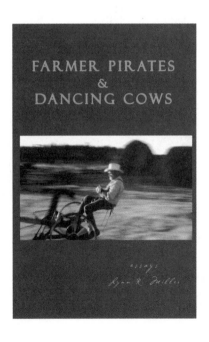

(Shipping $5 1st book [$7 other countries] $1.50 each additional)
To order contact; SFJ, P.O. Box 1627, Sisters, OR 97759
800-876-2893   www.smallfarmersjournal.com   541-549-2064

INDEX

Photo by Kristi Gilman-Miller

DISCARD

VERMONT STATE COLLEGES

0 0003 0845313 3

DATE DUE

NOV 18 2012

NOV 1 2 2012

DEC 1 4 2015

MAR 1 6 2017

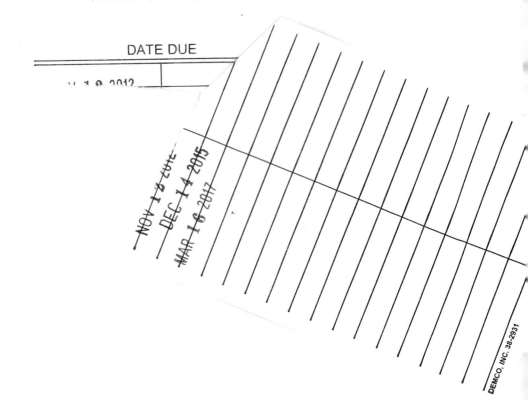

DEMCO, INC. 38-2931

Hartness Library System
Vermont Technical College
One Main Street
Randolph Center, VT 05061

DISCARD